Encyclopedia of Technology and science volume 1

Technology

Prof.Dr.Sami Al-Mudhaffar

Table of contents

Title	1
Table of contents	2
This book	7
Introduction	7
Chapter one Concepts in technology	16
Contents	17
Preface	18
The importance of technology	19
The technology concept	25
Technology and civilization	27
Technology and Heritage	30
Technology and philosophy	33
Technology and Development	42
Technology Transfer	44
Globalization and technology	53
International experiences in technology	**59**
Economy and Technology	63
Informational technology	64
Technological incubator	68
Chapter Two General Aspects of Technology	69
Contents	70
Preface	71
Technological planning	73
Technological reform	74
Foundations of technology	76
Novel challenges for technology	78
Global Experiences in Technology	82
Chapter Three Patterns of Technology	88
Contents	89
Preface	90

Technological option	91
Information technology in Iraq	93
The importance of technological incubators	102
Advanced Technologies	111
Chapter Four Technologies of analysis in clinical laboratories	118
Measurments techniques	119
Seperation technique	122
Characterization Techniques	127
Diagnostic Techniques	138
Nuclear techniques	141
Other Techniques	148
Chapter five Development of Biotechnology	180
Preface	181
Historical Aspects of Biotechnology in Iraq	182
Approaches of biotechnology in Iraq	184
Future biotechnology	190
The National Programme for the Biotechnology	196
Chapter six Technology of Education and higher education	204
Contents	205
Preface	205
The general features of the education system in Iraq	207
Technologies of Education	207
Contemporary technologies in education	209
The challenges of Iraqi education	211
Technologies of Scientific knowledge	212
Brain Drain and education	219
Global Experiences in Education	221
Futurism of education in Iraq	225

Toward the future of advanced education in Iraq	227
Visions of the future of education in Iraq	228
Renewal of education and missions and its role in development	230
Futures of the educational system	230
Priorities in the educational renewal	231
Iraqi educational philosophy	233
A look at the reality of education in Iraq	237
Technology and Higher Education	238
Global Tests in Higher Education	248
The challenges of Iraqi universities	250
Prospective of Higher Education in Iraq	251
New Vision and academic covenant	253
Relevance of higher education to national aspiration	256
The reality of Higher Education in Iraq	259
Facts about the development of the present of Higher Education in Iraq	261
Futurism of Higher Education in Iraq	262
Past and Current Status	266
Chapter seven Technological policies	268
Preface	269
Indicators and trends in technology polices in Iraq	273
Technology policies in Iraq	275
Proposal to develop a higher council for technology	**276**
Development of national policies for science and technology	278
Scientific and technological institutions in Iraq	282
Science and technology policies in some countries of the world And United States	283
Science and technology policies in the European Union	284
The relevant authorities in the formulation of world policies and technology	284
Proposed development of a higher council for science and technology	285

The Council of Science and Technology	294
Scientific and technological cooperation with Arab countries, foreign states and international organizations	299
Development of national policies for science and technology	304
Biotechnology policy in Iraq	320
Chapter eight Technological strategies	344
Entrance of the strategy	345
Strategic vision and future renewal of higher education in Iraq	346
New legal and academic vision of the strategy	347
Platform of the Strategic Plan for Higher Education	349
Higher education strategies	353
Building strategy principle for higher education in Iraq	358
Ways to implement the strategy	358
Strategic components of higher education in Iraq	360
Corner stones of the constitutional strategy	361
Aspects of strategic plan	362
Strategic plan for higher education in Iraq	364
Document of strategy of higher education in Iraq	365
Strategy for Education in Iraq	370
Ways and stages of implementation of the strategy for Education	371
The general framework of the strategy of education in Iraq	372
Draft strategy for the Iraqi education	375
Chpter nine Technology and science ,scientific research	386
Preface	387
Frontier sciences	389
Techniques used in science	**390**
Diagnostic Imaging	390
Protein engineering	395
Bio- engineering	401

References	440
Prof.Dr. Sami Al-Mudhaffar FIAS,IAS	443

This book

This book deals with the different types of technology and development in Iraq and the world. It deals with the updated concepts ,such as key issues and challenges facing technology and sector such as strategy adopted leading indicators and achievement, supervision, training and , planning and administration. It also looks at the technology that deals with frontier sciences and scientific research.

The book exposes the traditional functions of technology , such as; transfer of heritage. The inventory of the goals of technology in strengthening the economic feasibility as political systems is well covered by the book.

This technology , includes important issues that affect a large segment of people, with view to identify the reality of this type of technology, and then to develop a future vision for its advancement. This book shows how technology plays its natural role in developmental events in Iraq and make technology more sophisticated and in response to changes at the international, regional and local levels, as well as the possible extent, to be compatible with international standards.

The book is trying to monitor changes in an attempt to get Iraqi model, relates to perceptions of about new concepts of technology of global, local and regional and refers to the achievements that took place. The book also focuses on strategies for technology of different types, which have impact on the role of in preparing Iraqi rights to live in post-modernism or knowledge society, also represents a strategic vision for improving basis of challenges .

Introduction

The technology in general is characterized by different functions ,the philosophy goals and targets are usually derived from the national

heritage, the nature of governance, the economic and social system that prevails. The targets of technology vary according to the political philosophy that are managed by the specified country for example, the technological system in the United States of America is characterized by the spirit of hegemony, whereas the technological system of Japan is characterized by preparing the citizen by technology. Furthermore, the pragmatic technological system requires preparing a unique sound to the citizen in general, the goals sought to be achieved through its diverse establishments that include capabilities of researchers' social and cognitive skills, dissemination of technologist knowledge and preservation of the cultural heritage of the community. Furthermore, the technology has various functions that deal with developments such as, conducting studies, preparing the graduate with specifications such as; skills, scientific mind, and attributes of technological research for the purpose of competitions in labor market as well as teaching scientific research and community services.

The relationship between technology and the economy has taken root, when it led through global economic developments after the collapse of the former Soviet Union. The economic strategies oriented capitalist under several named including economic reform and corrective policies rely on market forces, and the management of the economy. Technology has become in the third millennium one of the strong economic growths and major venues for unemployment, brain drain, as well as other problems of financing the technology. Developmental performance of technology institutions depends on modern methods and quality assurance in harmony with rapid changes in the world, such as: the globalization of the economy and tremendous development of technologies .

Technology administration institution with legal personality, aimed at achieving progress in the areas of science, culture, art and thought, enrichment of human civilization, and expanding horizons of functional and human knowledge. The schools are also contributing to the achieved economic and social development, working to prepare specialists in various branches of knowledge and skills, participating in technical progress and development and pedagogical methods.

Iraq is facing constraints, such as, poor quality of information on technology and other stresses, and that the main source of information on technology in Iraq focuses on indicators of technology .Iraq society faces serious challenges related to the deteriorated situation, slow movement, and many heavy loads. It is unable to coordinate with the new society marked by knowledge and stun in its evolution in the technological accomplishment and productivity in different spheres of life. Furthermore, the quantity of technology output was reduced, which resulted in unemployment spreads, and the governmental institutions, refrained from recruitment of school graduates since those do not have skills and knowledge, which were required by work.

The technology is an model that reflects the social and economic context of the society, and an advanced stage of technology that provides the requested imagination with new vision. This type of technology is the featuring of the non-integrated image, due to the multiple economical and social variables that occurs in the developments of global communities.

The philosophy and goals of technology are usually derived from the nation's heritage, the nature of governance, and the economic and social system that prevails. The targets of technology vary according to the political philosophy, and they are managed by the specified country.

Technology seeks to accomplish the goals through its diverse establishments that include capabilities of technology social and cognitive skills, dissemination of technology knowledge, and preservation the cultural heritage of the community. Furthermore, the technology has various jobs that deal with developments, such as, conducting scientific studies, preparing the researches with specifications such as: skills, scientific mind, and attributes of science for the purpose of competition in labour market as well teaching scientific research and community services.

Universities are technologist, administrative institutions with legal personality and independence of financial management, aimed at achieving progress in the areas of technology, art and thought, enrich human civilization, expanding horizons of functional and human knowledge. The universities are also contributing to achieve economic and social development, working to prepare specialists in various branches of knowledge and skills and participating in technologic research and various studies that contribute to the technical progress and development of technology and pedagogical methods.

The university is a crowd of technologists and students, concerned with the duty of search for the truth; with this, the contemporary university is facing questioned phenomenon.

This technology in Iraq, includes important issues that affect a large segment of people, with view to identify the reality of this type of technology , and then to develop a future vision for its advancement. This book shows how technology plays its natural role in developmental events in Iraq and make technology more sophisticated and in response to changes at the international, regional and local levels, as well as the possible extent, to be compatible with international standards.

The book also focuses on strategies for technology , which have impact on the role of technology in preparing Iraqi rights to live in post-modernism or knowledge society, also represents a strategic vision for improving technology, on the basis of challenges and orientalism for a better future.

The book in your hands, deals with the content of the novel concepts of technology, culture, economy and education, technological leadership, concepts of technology , strategic planning and global experience in technology, as well as technology in the Arab world in many ways, including its establishment and reform. It also springs hope, that the display of the major achievements which are contained in this book, efforts reflect the sincere and tireless efforts towards the consolidation of the technological organization, to correct the path marked by a lot of violations and deviations in the past decades. The book in your hands deals in its content, novel concepts of culture, education, economy, the reform of technology, leadership, concepts of technology and strategic planning for technology, global experience in technology as well as in the Arab world in many ways, including its establishment and reform.

Given that we seek to reform radically the objectives, structures and contents and processes of technology events a quantum leap in output, taking into account that the quality of its various components is the heart of development in the reform.

The term technology refers to the application of knowledge in many areas and then described and studied and linked to different technologies.

The distinction between the technique and technology become a difficult, then both have been translated to ((technology)), in the early twentieth century by sociologists "Thorsten Vpelln" according to the German concept of the term "Technik" any the "Technology" expressed by sociologist Brian Reid including "all the tools and machinery, weapons, equipment and means of communication, transport equipment and skills".

Others researchers use technology, to expand the meaning by adding images and precision instruments, and Franklin said that the Europeans philosophers represent the same meaning as it expands different images.

The Ursula Franklin defined technology by "the real world in 1989" and presented also by the "definition and the way they operate the things around us , Bernard Staffler Varafha "to pursue life in different ways" and Mohammed Attia said the , "technology" is the systematic application of research design and appropriate scientific solutions and then developed to achieve specific objectives.

There are additional explanations depend on the specialization area, , and it consists of two syllables "Techno" which refers to the industry and "logy" which expresses the science where the latter is aimed at knowledge while technology "technology" is seeking to apply the objectives and programs they produce.

The updated developments indicate that technology represents the quality of human life in multiple areas, including the diagnosis of diseases and the development of means of transport and the second believes that technology, made problematic life suspension of intelligent and were not pre-existing instance, mentions.

The Hussein Bahaa Eddin said the world is experiencing a new technological revolution " so-called third that represents the information technology," "the technology of information which is the owner of social, political and cognitive control.

A common definition of the term technology focuses on the use of the calculator "computer" modern appliances, note that this view limited vision, since the computer is the product of technology , while the latter represents a way of thinking, problem solving and the use of knowledge, information and skills optimization of scientific knowledge and its applications and adapted to serve the rights and well-being.

Other researchers believe that the human materials and tools represent elements of the technology, and that the technological application begins the moment the interaction of these elements together, and Kausar Kojak define it as effort , human thought and the application of information, skills and availability that is needed and increase its capacity and the safety .

Other believes that technology, is organized application of the knowledge in particular field in order to obtain specific scientific results Hussein Kamel Bahaa El Din summarizes the vision to define the technology. "Technology" as thought and the performance of the solutions to the problems before just be the acquisition of equipment, and believed all of that technology, "technology" is not just the application of science or just devices, but human activity.

Contemporary Studies indicated and represents a quantum leap for human life, and introduced rights in serious problems were not exist before and some believe it leads to the destruction of human relationship to nature, where led to the emergence of new global culture beyond qualitative differences of the cultures of peoples and put an updated issues that represent globalization and human and spiritual values and other.

Perceptions and characteristics of the technology, according to the following "humanitarian scientific efforts , the modalities of thinking .experience in the use of information , expertise to solve the problems of human rights , satisfaction of their needs , increase their abilities and moreover it is applied science objectives and concept seeks to apply the knowledge, its inputs .

Organization, and output represents the final products that make up the complete system , the potential of technological consists of human resources "of skilled manpower and scientific talent capable", , the potential of technological innovation , scientific at the project level ,the organization of scientific and regulations, financial resources constitute the foundation stone in the technological potential to guide spending by setting priorities.

Many people called on the times we live in the era of technology, which include the use of all of the primitive tools and super-progress, as well as ancient and modern as the engine and machines that run energy , the automotive industry, radio, television , phone that affect the people and their customs and ways of life, and even ways to deal including working methods , technology of man overcoming hunger has facilitated disease prevention and treatment of many of them as a man enabled the transport of goods and passengers quickly.

This technologies brought about many problems including environmental pollution from car exhaust containing carbon monoxide gas and traffic jams, but the industries have managed to maintain the natural wealth of minerals and timber, not depleted by a process called re-industrialization name, retrieve raw materials of waste products ,their use in the manufacture of new products. Other examples that were used nuclear power plants to generate electricity characterized by several features outweigh the traditional electricity generation by power plants where the last act of burning fuel and emitting fumes and dust as well as that the fuel for nuclear plants to generate a huge amount of electricity using a very small amount of raw materials, but it caused huge amounts of hot water which flows into the lakes and streams has caused hurt thermally polluted water, animals and plants.

The contemporary technological revolution has changed the concepts of human reality in which they live and introduced new values fit with the current era and different culture collide with traditional culture and offers a typical life where qualitative differences between cultures disappeared and put the means and tools to produce one life and put a lot of abstract concepts that chime with tools that allow for the formation of human environment in which Art is, for example, give us a genuine answers to questions posed to human lives that there is a unification of the technology, "technology" and art in the current era.

Technology term have more than one definition, including the development and application of tools and the introduction of machinery, materials and automatic processes which help solve human problems resulting from human error and refers to all the ways that people use in their inventions and discoveries to meet their needs and satisfy their desires. The people through the ages have invented tools, machinery, materials, and techniques to make their work more accessible.

They also discovered a hydro power and electricity, which increased the work produced, the rate of modern communications, and data processing on this technology, especially technology of electronics .

The technology include the use of all of the primitive tools and high-progress and also the old methods of work and the research suggest that the man was able to make his tools for multiple purposes, such as manufacturing tools to be able to kill its victim and tools enable it to cut trees and other tools enables it to retain and storage and so on, it has developed from its capabilities in harnessing the energies of the animal, for example, transportation such as porters, transfer and manufacture of carts of these animals, for example.

We can say that such tools and methods that were used by the first man and was the target of various kinds, including the following:
- Facilitation and simplification his life with the things that he deals with.
- Increasing the production of food, clothing, supplied to a time of need.
- improvement of the quality of its products.

- Reliance on alternative energy sources for the physical abilities and to provide its requirements.
- Imposing its hegemony and control and project power on others.

Economically Karl Marx has defined the technology according to the following (members and tools achieved the production process) and (prostheses for the activity of social rights and a tool to control the order of nature) as well as socio-economic military phenomenon in essence, related to the culture and values of society prevailing, they are a set of experiences and knowledge and skills that must be provided to a particular industry or the production of several types of products in addition to the information that must be provided for the establishment of institutions and the establishment.

Finally, multiple views and opinions of writers and researchers on the concept of technology, including:
- The knowledge, experience gained, the physical, organizational tools and the means by which community needs to check the goods and services.
- The application of scientific laws and theories process in order to get certain needs.
- The application of results of modern scientific research in the fields of life process in order to satisfy certain needs and achieve specific objectives and influence in the environment to address a specific problem.
- Programming method for the application of scientific knowledge and experience in various fields of life programmer.

Chapter One

Concepts in technology

Contents

- Preface
- The importance of technology
- The technology concept
- Technology and civilization
- Technology and heritage
- Technology and philosophy
- Technology and development
- Technology and globalization
- International experiences in technology

Preface

In many of the terms of reference and investment financial resources in the development of specialized universities and cities and employment data technology to improve the type of products and reduce its cost, so it requires the adoption of a national policy for the development of technology that take into account local conditions. Technology is usually divided into traditional and advanced technology based on modern science and includes traditional technology, chemical industry, iron and steel and petrochemical industries.

The advanced technology is associated with advanced science of new materials at high temperatures such as microelectronics technology, computational designs, applied technology of robot, technology of laser, optical fiber and biotechnology. There are technological gaps from country to country depending on the size of the technology prevailing in the economy, which vary according to the theory of internal growth depending on the route of the technological path. The size of the technological gap between the different countries, are effected by determinants including the size of the technology and the possibilities of innovation in the economy and the flow of technology transfer.

The term technology transfer indicate that technology is gained through learning by doing, not just through the transfer or importation of goods or services refers technological and therefore they are not the manufacture of pre-prepared removable and cause technological deficit.

The experience of Iraq in technology transfer are diverse, including the large margin completed with foreign companies, which provided a comprehensive and complex technological transactions in the framework of the international market strategy. Iraq has suffered from indiscriminate transfers that took place in the absence of any sound domestic policy to create an independent local base in various fields of technology.

Thus, Iraq is facing two problems related to technology transfer the first is the search for modern technology and its transition and assimilation, development and improvement and the second related to the development of the technology.

Accordingly Iraq needs in the field of technology transfer the following :
. Searching for all technological alternatives.
. Selection of appropriate technology.
. Adaptation of the selected technology.
. Determination of the problems of the adaptation of modern technology.

The importance of technology

Some thinks that the technology are divided into two concepts the first believes that technology represents a quantum leap for human life, and refers to a large extent to the elimination of many of the constraints of human life, such as the diagnosis of disease and the speed of transport, and the second holds the modern technology that enter human in serious problems were not existed before and this concept focuses on the critique of technology, and that technology has contributed to human molding and the formulation of his life in the image were not existed before.

Despite a recognition of researchers of the importance of technology and its role in human life and its ability to contribute to the realization of justice, freedom and some believe it leads to the destruction of a relationship of human and nature.

The contemporary technological revolution has changed the human concepts of reality s, because contemporary society can not live in the middle ages or the values that emerged from the Industrial revolution to the recent values. The introduction of new values fit in with the current era, so contemporary technology herald the values and culture different collide with local culture because it provides tools and means to produce similar in dealing with community issues. The technology led to the emergence of a new global culture beyond the qualitative differences of the cultures of peoples and has put these concepts issues such as globalization, human and spiritual values, and others.

The technology transfer the scientific results to a variety of tools ,and knowledge , art investment represents a creative capacity, as well as inventions and applications, and Karl Marx was the first researchers economists who fired term technology at the beginning of the nineteenth century.

Economically Karl Marx has defined technology as (members and tools achieved the production process) and (prostheses for the activity of social rights and a tool to control the order of nature) as well as socio-economic phenomenon military in essence, related to the culture and values of society prevailing, they are a set of experiences and knowledge and skills that must be provided to a particular industry or the production of several types of products in addition to the information that must be provided for the establishment of institutions and the establishment of a very mentioned.

Finally, multiple views and opinions of writers and researchers on the concept of technology, including:
- The group of knowledge , gained experience ,the physical , organizational tools and the means by which community needed to check the goods and services group.

- The application of technological laws and theories process in order to get certain needs.
- The results of the application of modern scientific research in the fields of life process in order to satisfy certain needs and achieve specific objectives and influence in the environment to address a specific problem.

Technological challenges
The following points represent the most important of these challenges:
  ☐ The complexity of human and technological resources between companies.

- Exchange of information and transform information into knowledge.
- Restructuring of companies and institutions.
- Participatory between economic and social actors.
- Providing general legislative and institutional environment.
- The deterioration of the security situation.
- Resistance to change, in strategic planning.
- Response of institutions to the technological initiatives at hand.
- The private sector response to contribute to the investment in scientific research.

A favorable contemporary studies of technology indicated a difference in the level of interest , in the technological progress and mechanisms associated with it within both developing and developed countries as well as different in other economic factors in developing countries . The differences in the mechanisms associated with the technological progress between developing and developed countries ,updated information and refers to the difference in interest of developed and developing countries the process of technological progress.

The developed countries of pioneering technology through innovations , research and development in all fields, especially on the part of companies and government and non-governmental organizations and institutions within the state are also interested in developing countries.

The process of technological progress require the basic needs within the State and can be different mechanisms adoption in technological progress represented by the reasons of the pursuit of technological progress and ways to achieve technological progress and sources of technological progress and the availability of the necessary resources to achieve technological progress and the work environment affecting the technological progress

The modern technology is facing the challenges of no limits, the first of these challenges is the overcoming the bad side effects of this technique through the introduction of new technologies and then developing it and there is still another challenge is the distribution of the benefits of these technologies and the virtues of the world, particularly the developing world.

There are difficulties in many cases to overcome some of the negative effects of the technique, as well as the treatment each other or get rid of it. In spite of these disadvantages, it can be overcome once again the problem of unemployment, through the cooperation between the owners of industry and government, to re-train these workers and then to fill jobs that require a higher degree of skill, it is likely that the new jobs be more acceptable and convinced.

Those in charge of the industry can work a lot to reduce pollution of the environment resulting from the problems of different industries, as well as reducing the depletion of natural resources. The development of alternative technologies, including the possibility of overcoming the problem of air pollution to reach a sophisticated technical means to purify emittion from car exhaust gases .

Nuclear fuel stations also generate a tremendous amount of electricity using very small amount of raw materials, and they do not pollute the air, as is the case in existing power stations to burn fuel that flow huge amounts of hot water which flows into the lakes and streams. Scientists and engineers are currently working to solve this problem in nuclear power plants, the establishment of users of air cooling towers to cool the hot water resulting from the stations. Technology challenges, which was launched since the end of World War II and then accelerated and constitutes a terrible results in the last few years in the wake of what is known as a crisis of energy. The original technology revolution branched out recently to a group of scientific and technological revolutions in order to deal with the concept of challenge of the technology that requires a grasp of science concepts and technology , economic growth and development process , the problems of technology transfer and analyzed scientifically . In addition, the study of the causes of the scientific backwardness at present and the past are almost a consensus at the present time among scientists that the technological advances constitutes one of the most important Factors for economic growth, if not the most important at all.

The increase in the average per capita income is no longer only to labor and capital, but other factors remaining incorporate them by economists under the umbrella of the technology and found from different studies in different periods that technological advances have dominant role in the increase of the rate growth. Japan has faced challenges, including the Western and the US challenge in the last century and in its middle called constructive shocked and shock of Hiroshima and Nagasaki bombings hit led to the Japanese to overcome the technological gap that separates Japan from the West and get the technology and control it . The challenge of the control of Japan on modern technology and its success in creating technology is solid and environment research, especially in the industry led to a very important development on the overall economic activity in Japan and rising labor productivity component and relatively low cost of labor.

Appropriate technology can figure out the appropriate technology through its definition and standards, including: - appropriate engineering the entire technical and organizational methods and machinery, equipment and speed of completion of the work and the quality of the methods and machinery, equipment and suffer not only from the inadequacy of the technology imported by local and geographical environment but also of important negative effects created by developing countries some imported technologies to lend themselves with the social and economic conditions prevailing in most developing countries.

The importance of technology and its role in human life and its ability to contribute to the achievement of justice and freedom, recognition and some believe leads to the destruction of human relationship to nature and affect human relationship itself. The contemporary technology revolution has changed the human concepts of reality in their lives, or the values that emerged from the industrial revolution to the values, but is, the introduction of new values fit in with the current era of ethics of technology.

The technology take the recipe challenge of the human race , social variables and the impact of these technologies on the humanitarian , human rights , moral values , scientists and researchers will be required to make their achievements in the form of knowledge formulated to accurately shared by people.

It is clear that both genetics in its original form, biotechnology by multiple of technologies represented by framework of the signs of the natural evolution, of Darwin and called for Spencer in social image and faced sharp criticism in the late nineteenth century. The moral repercussions of the technologies with a distinct interest in enriching the scientific research, which is still growing and evolving as raises serious questions together constitute the social issues and scientific issues.

The religions confirmed on ethics of researcher and research and both are two sides of the same coin .The search must be heading for reconstruction , the development and preservation of the environment. The treatment by the human gene , infertility and eugenics need to research and study and will remain controversial among different currents and contradictory beliefs which is reflected negatively on our future generations.

The gene therapy in somatic cells treat some diseases morally acceptable but gene therapy in stem cells remains a subject of controversy with regard to the treatment of structural cells. The use of viral vectors in human gene therapy is a subject of great interest due to the ability of these vectors to make a satisfactory particles may spread to neighboring cells and to the other people in the community.

There are potential risks in the use of gene transfer by retrogressive virus causing lymphocyte malignancy in T cells of monkeys, but discovered later that this is due to the carrier virus and may not hold any search contamination or perform any treatment or diagnosis of respect genome of person only after a rigorous assessment and a precondition. The potential benefits associated with these may not be any research on the human genome , genetics and medicine that the commitment to human rights and human freedoms any individual or group of individuals.

Technologist believe that their goal is the service of humanity through the discovery and development of products to the process, as is the case in, and car technology, television and the application of genetics, cloning and research and technology. Historians does not know their part for the technology and tools of social aspects and impacts of humanity such as robots ,other tools and machinery so requires awareness and understanding of them, people and society from the other side to create a suitable ground to understand the technology requirements and to find the relationship between technology and the substantive issues of society.

The technology concept

The technology concept transfer the scientific results to a variety of tools and knowledge and art investment, represents a creative capacity, as well as inventions and applications. Technology has helped human groups to overcome nature, and then provide the style of civilized life.

The first man, did not have the means to control nature, but only a modest primitive tools, and was ignorant of animal husbandry, did not know any style of agriculture, and with the passage of time, then the man discovered how firing kindles. This discovery has helped man to better control the circumstances surrounding it, also known as the methods of crop production.

The development of agriculture and the production of various agricultural crops to the stability of rights in certain locations, and then building human societies, which gave freedom and the time to do the other side of food production.

Industrial technology began the discovery of the steam engine and the machines run by energy, and that the evolution of the auto industry affected the people greatly , those who work in factories and living near them, the radio and television have changed the people's habits and lifestyles and even in ways of dealing with them. The discovery of phone created a large revolution in the means of communication.

The technology has made it possible for humans the means for overcoming hunger and facilitated the treatment a lot of diseases, as enabled man the transport of goods and passengers quickly to anywhere on the globe, and enabled the technology of man out of the earth and walk around in outer space and set foot on lunar surface.

The transformation of the industrialized nations to the use of agricultural tractors, and other machines working with oil or electricity on farms, to increase agricultural production accompanied by the use of agricultural machinery, fertilization use and methods of agricultural technology developed huge increase in production, for example, most of the workers of factories in the beginning of the nineteenth century, completed a manual manner or manual bales, but today, it has solved-powered machinery replaced manual labor to a large extent.

Thus, the advanced industrial countries provided for its citizens much better living standards than the non-industrial. Citizens of industrialized nations fed and clothed in best manner, and better homes provided with all services. Human in these countries enjoy healthy life free of disease, they are more well-being than any other people found throughout history.

Globally, technological revolution spawned other enormous positive traces at various levels practical and theoretical, and negative results are still the subject of debate. Any urban progress may reflect negatively due to increased breadth of the cultural gap, it is also noted that the intellectual power, including science and knowledge, has become a prior. Adam Smith (A.SMITH) have realized this fact and has been noted that human skills and creativity are more important Than raw materials. Arnold Toynbee (A.TOYNBEE) reiterates famous sayings, that the third world in general, will not be able to cope and face the west or unless use the same tools through the development of social and economic conditions, the adoption of advanced technologies able to achieve steady progress, as happened in Europe.

The Orientalist Maxime Rodinson says the lack of development of knowledge tools (concepts, methods, vision ...) in the Arab and Islamic culture makes them unable to accomplish the rise of intellectual and practical steady progress, and its evolution of civilization, scientific, technological and economic, and the failure of Arab countries behind refers to: -

- Failure of the role of the mind and technological community in the Arab countries.
- The Arab mind slowed in its passive role in the development of civilization crisis,.
- The lack of solutions to overcome the crisis of the development of civilization.
- Survival of the Arab and Islamic scientific heritage away from the current cultural awareness, while the religious and literary heritage occupied the full care and attention.

Technology and civilization

Fouad Zakaria said that the technology represents tools and methods that are used for the purposes of the current revolutions in the community in which they can draw a picture of a bright future of humanity if it worked in accordance with the moral and social concepts. Many researchers have predicted the most important cultural, technological advances that have been made, including the IT and the Internet, telecommunications and artificial intelligence, space travel and Biotechnology and Genetic Engineering.

The largest and most important advantage in information technology and new communications bypassed and ignored the language and local culture and traditions barriers, either with regard to biotechnology and genetic engineering (genetic engineering) of course is completely different, the world has witnessed a rapid and surprising progress since developed the reproduction of sheep dolly, and then human genome project become at the forefront current scientific concerns. The high technology is represented by the Trinity composed of computer knowledge ,genetic engineering and nanotechnology . Some believe in the prospect of human cloning enormous results, possibilities and piercing some of them positive and some serious in social and cultural level.

Civilization is the result of human interaction with both nature and society, and technology is part of the toll of human interaction with nature, they are part of the solutions that become as soon as they occur excellent quality for both, they are not making history, but provide elements of workmanship.

In the past Ibn Khaldun used the term "construction" to denote and witnessed the level of people in their life and the promise of civilization as Urbanism and its limit. In the stage of civilization people are, according to the opinion of Ibn Khaldun, they reached the redundant case . "Will Durant" defined civilization as social system assigns to help people and Ratnao (Rathnau) , Xeirlnj (Keyserling) to the increase of their cultural production necessary of the conditions of livelihood. Thomas Mann (Tomas mann) and other german researchers expressed the meaning of civilization as deep spirit ofcommunity based on the confirmation of spiritual authenticity and truth philosophy of emotional human being spirit man. McKeever (Maciver) said, the civilization, is represented by the arts, literature , religions and ethics .

Edward b. Taylor the English sociologist has said that the civilization of any society is a complex consisting of all knowledge and belief, art, morals, law and traditions.

From the above, we find researchers divided in the meaning of the word civilization in two teams:

- Team sees civilization means of a group of intellectual and physical manifestations in society.

- Other teams sees civilization as intellectual appearances only.

The theories revolving around the emergence of civilization are the following:

Enviromental theory

A theory go to emphasize the role of environment factor in the emergence of civilization, and this theory emerged by Greece thinkers . Ibn Khaldun has been influenced in his introduction by the theory of the environment, recalled the impact of the environment on human skin blacker and whiter, and on the eyes bluer and bluer is, and on the body in color purity.

The opinion of Ibn Khaldun:

The occurrence of the Ibn Khaldun under the influence of the views of Greece thinkers to confirm the role of the environment in the civilization elsewhere in the preamble explanation for the emergence of the state in people's lives, Ibn Khaldun that humans meeting sees the need dictated by the requirements of life was to be for human beings of cooperation which is the basis Human meeting followed by the inauguration of the Working mostly the same arbiter in the affairs of people and their issues.

- FICO theory:

The Italian philosopher Jean Battista Vico (1668-1744 AD) study of mythological traditions , popular mythology and ancient poems like poems of Homer and primitive legislation such as the twelve panels legislation, and finished to the development of the theory of "historical cycles" in his book "The new science and the principles of the philosophy of history." FICO tried through his theory to show the laws needed to study the fate of the peoples of its inception, progress and advancement to its decline and the end .Then he said, the laws should be one to study the formation of peoples .

- Sex, race theory:

The term sex uses to express the availability of some distinctive traits inherited in certain groups of people, and the theory of sex shows the color as the owners of physical capacity is more reliable than others.

- Spengler theory:

The German philosopher Spengler "1880 -1936" believes that civilizations are historical objects behave like living organic organisms, have one context going for it and the similar cycle pass by. The birth of civilizations and its development , Spengler says, "breeding civilizations started at the moment of awakening up in large spirit of separated from the spiritual preliminary condition of children's eternal humanity, as inseparable from what the picture is not its picture. The civilization die when the soul of civilization has achieved what all of the potential and languages.

- The bias theory and response:

Arnold Toynbee thinks the existence factor in the emergence of civilization is not a single but multiple, with joint relationship that is when human exposure to the challenge and then its response. Challenge and response are closest positive factor according to Toynbee explaination of the emergence of civilization and the reduction of the tender rights in the case of infertility in another case.

- Materialization ,Physical theory:

This theory include the self-developed evolution of materials for creation explanation excluding the excistence of creator .The theory also explain that the society is a part of physical theory.

Technology and Heritage

The word "heritage" is derived from the verb "inherited" but is included in the modern era, new meanings loaded with emotional shipments and religious implications may exceed the word's English heritage or tradition derived from the Latin word tradition and the term "heritage" indicates today's cultural, intellectual, religious, literary and artistic heritage and technological. Mohammed Abed al-Jabri said Heritage is not perceived it as one of the past but it as "complementation of culture" and adds, "The heritage in the modern arab consciousness does not mean only holds possibilities that have been made.

Some of the secrets of heritage development and technological ensure, for example, billions of galaxies and black holes and the worlds bizarre and nuclei of living cells and the code of life and the promotion of human knowledge and new prospects in various fields of life. The evolution of the technology resulted in findings of technological beginning of multi-power benefits, to use satellites to be addressed diseases, and transplantation of human organs genetic engineering, and bridge the geographical and intellectual distance between human beings, and all this leads to interaction and cross-fertilization of cultures, civilizations.

There is no doubt that the relationship between technology and culture among the topics that have attracted and still attract the attention of researchers to elucidate the reservoirs this dialectical relationship between material and spiritual, and in this context, important study is required to add a brick to the previous ideas dissecting properly for this relationship.

The technology associated with the manufacture of ancient civilizations, including the civilization of Mesopotamia and the Nile civilization of the Greeks, and thus represents the outcome of different civilizations and not limited to a specific culture in the modern civilization A distinctive technological signs, including space exploration and biotechnology and genetic engineering and the Internet and nanotechnology.

Stone age industry tools and instruments of stone and dating back to hundreds of thousands of past years, followed by the Stone Age, the Bronze Age, which include adoption of bronze instead of the material that was made from mans tools, a stone which was able to is create in their uses and adapted for the production of tools that meet their needs and contribute to its progress and development. The era of the evolution of the technology industry where it became the iron that produces tools for different fields of life and it covers the fields of agriculture, industry and others, which became known as the period of industrial revolution.

Britain turned in the period from the mid-sixteenth century to the early seventeenth century from an agricultural society and then rolled inventions and discoveries in the past of three centuries extension to form what became known as the industrial Revolution.

On the other hand inventions of the nineteenth century and the twentieth century contributed to the invention of the radio, vehicles, aircraft and others in the community development and growth and prosperity. On the other hand developed societies shifted from reliance on mechanization (machines) to the mechanism of production and then the information and communications revolution came in a very short time. Of the greatest technological innovations carried out and made to quench the ground and facilitate the modalities of transportation and communication between the various parts of the country, while the ancient Greeks through Greek mathematics, Egyptian to technology or Sumerian, was developed rotational movement applied in the mills and pistons aerobic and amplifiers.

Arabs were not content with the transfer of Greek heritage, but they adopted the Persian and Indian heritage, shown in the innovative compositions ranging from 399 e and Alkrgi 410 e and tents 515 e and Nasir al-Din Tusi 672 E, ,does not accept the argument that the task of the Arab Muslim scholars were not rely on experience alone, but the theory is preceded by verification of the theory. The boom mechanism Arts in the Arab world and the Islamic world, in the peak period between the third and sixth AH / ninth and the third century AD, ten of technology that has been established in the Near and Middle East in many centuries.

Egyptians and Greeks, Romans, Byzantines paid in different ways of heritage strides forward and these machines are designed for use daily and water pumping machines and other . The machinery has been transferred to the west or road, the technology boom, by the sons of Moses and 259 e 430 e Ibn al-Haytham Ismail island and Taqi al-Din observer 993 e.

In the time of the Abbasid background safe, 218 e, and in Andalus represented by Abbas Ibn Firnas, 274 e and trying to fly. The technologies mentioned and others involved utmost importance, since much of modern technology , those machines consummate, especially micro-technology and medical devices.

After World War II, as a result of the disparity gap between developed and developing countries, scientists using the new means of radio and television conduits media to help the bridge that gap and flocked to developing countries, between the two Arab countries to use the new technologies of radio, television and cinema, or the strengthening of what was available to have them despite the fact that the, the jump of development to catch up with the countries of the developed world in a short period did not materialize .

As the results of research references to deepen the gap of and increasing widening between developed and developing countries on the other hand. Among many observers of Technology from a historical perspective of three revolutions of technology began the era of mechanization, and passed the era of automation and moved into the era of high-end technology or as they say the technology superior and of the last of the twentieth century three decades and the first decade of the atheist and the twentieth century and which we can call it technological revolution, the third industrial revolution, the first representing the supply of human potential muscle and mentality represented by the cranes and machine tools indicate the second industrial revolution to exempt rights values of chores refined and is the third for economic, military and political monopoly of the States legibility and its components sovereignty of the Industrial Revolution. The technological revolution are characterized by technological events that are coming in surprisingly quickly.

Technology and philosophy

Technology linked to the philosophy that provide with ideas and previous concepts, and dealt with a critical rational ideas and Plato takes mental and metaphysical meditation isolated from objective reality perceived . Aristotle separate mental consideration n and its practical application, in the field of knowledge it separates the theoretical mind, which deals with the essence of the idea and the practical mind which deals with the material .

In modern western thought Francis Bacon, who started from the mind criticism, he find a new mind must be configured based on a new logic instead of the old mind based on the logic of Aristotle, who moved away from the experience and knowledge of what being targeted.

The term "Philosophy" was originally derived from Greeks, composed of two parts (Fib - Villa) with the meaning of loving and word (Sophie - Sofia), which means wisdom. In the Arab-Islamic heritage it had several expressions such as morality. Specification of the most important fields of philosophy are a broad and many kinds of sequential steps leading to the formation of thought to simplify the basic perceptions and understanding of human dilemmas. But they are all looking for the nature of things using the mind. The philosophy was spreaded in Greece with idealism of a lot of contradictory intellectual movements, but more of idealism and realism the view is better for the values and virtues constant that does not change while the second believes that care of the most important senses is better than to focus on the imagination and consistent with the ideal of the virtues that fixed.

It is intended that basics of philosophical application of theory in the field of technology for the purpose of clarifying the technological process and to identify its features, which is also being organized as intellectual activity, depends on the philosophy as a tool to organize and coordinate the technological process. Technological philosophy is required due to technology as a part of human existence totally as well as, science and language.

The concept of technology differ according to the kind of philosophy that deals with it, indicates that education is the formulation of the human being announced and well-being, and Plato believed that education is the consistency between the self and body, while Abu Hamid Al- Ghazali believed that technology is a priority for humanitarian and spiritual atmosphere. Education leads to higher degree of ideal maturity for children, and natural philosophy refers to the mental preparation of the child. While existential philosophy believes that man is free and subject to the inevitable and philosophy of pragmatism which is the vision of Dewey that suggests that education is a permanent organization of experience and adaption to social reality.

John Dewey, one of the philosophers, who pointed the principles underlying the concept of modern technology , including that technology is a small community that comes to life and education continues forever, and the curriculum must be accompanied by life and mission of education is to prepare the individual for life.

Among the Arab philosophers, who were interested in the problems of education, are lbn Sina, Al-Ghazali and lbn Khaldoun (1332-1406 AD) whose belief was limited to the educational principles characterized by gradual transition from known to the unknown, and from easy to difficult.

The European educationalists, such as: Jean-Jacques Rousseau (1712-1778 AD) who called for equality and return to normal life away from corruption, and who called for meeting the needs of children, and Herbert Spencer (1820-1903 AD), who focused on the psychological concept of special education and child psychology.

The subject of philosophy is knowledge of the natural facts (relative) standard and the facts (values and ethics) and the philosophy is considered as perceived by Dewey. For example, the general theory of technology is characterized by a turning humanity of human beings to the intellectual and others believe the philosophy Celebrities are the leaders of technology thought Ksagrat, Confucius, Plato, Jean-Jacques Rousseau, which reflect about the conditions of philosophy of technology and work to evaluate and critique the educational process. As the fields of philosophy and issues of our time metaphysics (metaphysics), or the divine science, knowledge, values and philosophy of education is to apply the positivist approach in the field of human experience which we call education.

In other words, the view is emanating from the educational theories and philosophical ideas within the framework of cultural and technological philosophy that does not consider the method of abstract ideas, but only consider how to use technological philosophy us to devise channels, principles and foundations required for work in education, educational teaching practice, learning and human development. It shows signs of Greek civilization which includes the Investigation to examine the extensive philosophical basis of human behavior.

It is linked to the philosophy of Plato, which is piloted by the perception of two worlds, ideals (fixed) and the real world (variable). The community is composed of two layers, one of them think and other works in other words, the first class is linked to the educational framework for the purpose of access to knowledge, which requires the mind, and thus knowledge is a real constant and does not change. The philosophy believes in and basic principles, based on the ideas of absolute belief in the existence of independent real in a perfect world.

This philosophical school follows fixed doctrine and the development of the idea and the teaching methods whereupon the basis of mental training of personnel. The features of application of this philosophy in the field of education characterized by the accumulation of clear knowledge approach, and the adoption of fixed educational tools, routes teaching examinations without the use of traditional means. As well as that of Plato and tutor Ferry Socrates (470 BC - 399 BC), show fixed values and virtues does not change, the technology should take care of reflection and imagination, and that human nature is composed of two parts, soul and the body, and must take into account this duality for the upbringing of the individual in society as deviant.

According to these realistic scenarios, technology then must transcend the soul without neglecting the body. The ideal philosophy of technology then introduced Socratic generation idea which is based on the mind provoked to discuss the self-addition. Furthermore, the ideal technology tends to focus on philosophical issues with the experimental spirit with abstract thinking and archiving of information exercised by school.

According to this philosophy of Aristotle (384 BC - 322 BC), that there is only one world, that is the real world, characterized by the firm principles and the care of the senses. It is more important than a focus on imagination and consistent with the ideal of the virtues that is fixed. These realistic narrators follows stability approach and this philosophy depends on the continuous attempts to discover the universe and the world and work to understand the existing laws that include all the facts within a stable and consistent world. The, duties placed on the philosophy of realism for Education illustrated as follows:

- Following the approach to accumulate the knowledge of the natural, social and cultural rights.
- Acquiring the knowledge, skills and habits to prepare students for life.
- Scientists preparation with distinguished scientific criterion.

- Extra-curricular activities are essential to the realist school.
- Extra-curricular activities are an important part of the technology.
- Curriculum consists of a set of facts discovered by scientists.
- The use of programmed technology machines.

The fact that the realist school has raised the status of the senses dropped the philosophy from meditation, the ideals and reality, imagination to the senses, thus arena of scientific knowledge and curricula astronomical sciences. John Locke (1632 - 1704 AD), is a pioneer of realism philosophy because it was backed by conviction and gave critical thinking, a wide range of experience and reality and the senses example, not based on abstract knowledge and science. This philosophy view that the child is a blank sheet of the lines from reality, as well as Locke had added the need to study natural phenomena as well as mathematics and other sciences.

The Komenus (1592 - 1670) a realistic perception was making the image of the most important methods of teaching children in schools along with the realities of enthusiasm and believed to be the first founder of the technological freedom for children. And he developed a sensory method of teaching and took care of the physical and moral education to both.

Finally, the factual findings to multiple convictions, including:
- Establishment a method of experimentation and exchange of scientific and systematic uncertainty).
- Encouragement the technologist to observe natural phenomena in an order system.
- It was called for acts of reason in the analysis and the independence of the senses.
- The technology did not distinguish between the world of ideals and spirit and the serious and actual world.
- Elaboration on intellectual reflection.
- Focusing on professional technology.

The roots of this philosophy belong to ancient times and the writings of Hrikulais (535 to 475 BC). But Undtlian (30 to 95 ac) and Harzubayrs and

William James (1842 to 1910). While contemporary pragmatism associated with the new world (the second half of the nineteenth century and early twentieth century), called pragmatic, instrumental, developmental and operational.

Of the principles of this philosophy, education is life and this philosophy of education linked to community service while giving the technologist a degree of freedom. And human beings adapt to the biological environment, the world is relatively constant and constantly changing, the truth is not absolute but changing society and democratic decision making.

Pragmatic philosophy plays key role in the development of technology methods and improving the traditional method and manner of trial and error. The school of John Dewey is based on the philosophy of pragmatism played historical intellectual revolution against traditional technology , which focused on information and not believes in the values of morality, but according to the relative renewed convictions of society.

The technologist in the philosophy of pragmatism does not teach the traditional materials in a systematic way, the technologist moves from one idea to another in a sequential manner and deal with each idea on the grounds that suggests to his students the problems of the future. And that the values vary depending on the situation and the duration of the community and individual's convictions.

The technologist in the pragmatic philosophy is not a process of transmitting knowledge to students for knowledge, but help them to meet the needs of the social environment and able to raise the forces required by the student, including social attitudes. Perceptions of the pragmatic philosophy depends on building curriculum writing arithmetic means rather than targets, and approach interests in the facts relating to the nature of the child. Pragmatic philosophy stresses on the development of vocational education, natural sciences, while the humanities and languages is of secondary importance.

This philosophy has paved the way to progressive education "Progression" and that some consider it as the application to education. The most prominent pioneers in this respect is John Dewey at the beginning of the twentieth century (1920 - 1945 AD), which came as a revolution due to the disadvantages of the traditional system and excreted key concepts including:

- Coordination with the environment as the best climate for technology
- Take into account the individual differences of each researcher
- The student has to ask about the environment surrounding
- School has to insulate from the external environment and adopt the way of solving problems
- The values and technology is relatively changeable.

This philosophy believes that man is good by nature and what is the reason for corruption of man is due to the society and its institutions. Technology has a target of opportunity for the natural growth. This philosophy also believes that the sensors are sources of technology and entry points for the development of thought and not the role of the stock of knowledge. The natural philosophy calls for the reference to the technological activities that are consistent with the original natural Laws. Rousseau (1712 - 1778) formulate the ideas of natural philosophy, the role of nature in the development of technology is illustrated in terms of the needs of the technology and the need for tilting the obstacles facing it, and in accordance with environmental requirements.

Existential philosophy "Existentialism" and the Existentialists Deckard and Sartre believed in the seriousness and absolute human freedom. It is a philosophical vision of human existence, responsibility and human nature and the world, knowledge and existential values emerged in Europe after the First World War (1914 - 1918) in Germany and later in France. It is believed in technology as perceived by them and that man does not accept the existential indoctrination and technology to promote the existence of rights, and literature, music, philosophy and the arts are of the requirements of public dialogue and debate.

The pioneers of modern existentialist such as Jean-Paul Satire (1905 - 1980) believes in the following:
- The religion and constant values are not weighted.
- Public, call for the legalization of sex activities
- Ridiculing religions.
- Reverence for human beings.
- The existential deny the use of punishment and rebuke.

Some believes that the old Egyptians are the top source of technology. Egyptian civilization that is manufactured Hellenic civilization and technology was often based on temples (Am Shams and the Temple of Karnak). The teaching was done on wood, ebony, ivory, and papyrus sheets of paper. Others believed that the concept of education in the context of the contemporary science that has been by the ancient Egyptians, and teaching aids had been started before them by the ancient Egyptian drawings carved on the front walls of temples as a means to connect the sensory information. The ancient Egyptians also knew as public libraries.

The Greek history was represented by two cities (Sparta and Athens) have agreed to make education a means to prepare the citizen who served his country, in Sparta educational strategy was focused on the ongoing military exercises with distinguished scientists such as achievements of the father of medicine. Hippocrates discussed the medical issues in theory and practice. "Hippocratic oath" of the oldest are still the rare historical documents that show the practice of ethics of the medical profession. The Greeks were the first people who began to study memory and functions of the brain and heart.

The Romanian culture has played a role in the history of human thinking, where education was focused on the skill of reading, writing and learning in small schools, primary schools were known as Allodoux called Palmed teacher and educational ideas still protruding as educational as guidance.

Chinese civilization is represented in many sciences, including medicine, chemistry, and the Chinese needles are still, means of treatment.

China has Eastern culture that is rigorous, objective assessment of the tradition and the past and commitment to a series of teachings and provision of humanitarian tradition and appreciation of the past and focused on:
- Hold the tests for students.
- Save the information.
- Compliance with laws and legislation.

Ava Confucius (479 BC.-455 BC.): One of the foundations of the Alkonfohicep movement, where it used several means of socialization, includes:
- Take advantage of political office.
- Caring for skillfully reading books.
- Study of language and literature.
- Subordination to the father.
- Obedience the ruler.

Therefore, the Chinese education is characterized by:
- Ethics education.
- Clarity of vision.
- Faith in continuing education.
- Focus on the heritage of the former.

Hindi Civilization is various, such as, of Buddhism, which is characterized by full-time self-worship and the liquidation of all the things that surrounded the difficulties and desires of the causes of suffering. The Buddha's teachings have won wide acclaim in the world, its educational profiles was the duty to promote education and teaching by Interviewing and the lecture. Buddhism concentrated commandments include not to murder, adultery, theft and alcohol. Hindi civilization philosophy includes study of medicine, stars, science, and mathematics.

The cultural anecdotes shown in Kalila and Dimna care about the reform of the soul according to the Indian philosopher Alipidba. The educational features of Hindi are represented by the spiritual and moral education and social development.

Western Renaissance began in the fourteenth century. After that the reform era began in the sixteenth century, where the idea then matured compulsory education at the primary level which indicates:

- Focus on the concept of the importance of education for children and religious education by John Locke (1632 to 1704).
- Studies of Jean-Jacques Rousseau (1712 -1778 m) on the nature of human rights and the role of society in its composition.
- Building a serious mental test in Western culture by Francis Nikon (1561 - 1626) where the rejection of the traditional intellectual approaches that has been done in the middle ages.
- Development the process of teacher preparation by John Frederick the founder of modern pedagogy in the nineteenth century (1776 - 1841 AD) that focus on the importance of diction and methods of linking information, ancient and modern, and others.
- Provision of free primary education and its necessity by the horse, as well as the preparation of teachers in the thirties of the nineteenth century.

In the twentieth century emerged Dewey and put forward his idea for a lot of things, teaching and education as the role of schools in the community and linking the philosophy and education closely linked, as well as Pavlov and Skinner, who focused on the importance of promoting positive behavior reward.

Technology and Development

Economic development for example, is represented by the transition from agricultural economy to an industrial economy based on the acquisition of factories and means of production policy, but the limitation the acquisition of manufacturing of production means made policies around aging technologies to enterprises uncompetitive.

The economic boom of trade and encouragement of investment would have created an environment conducive to the transmission and resettlement of technology better than the previous method based on the purchase of factories and means of production. The relationship between technological progress and rates of economic and social development has become stronger than ever, and then the investment in R & D sector of technology has achieved the highest overall return on investment.

The technologies of knowledge the most important element in the economic development process and estimated its economists now that more than 50% of gross domestic product is based on knowledge, reaching built the proportion of exports to the knowledge and in the United States to 37% in Japan, 36% in Ireland, 43% in Britain, 32% . That access to knowledge is the way to make a quantum leap in the life of communities, so the idea of knowledge societies become an attractive idea depends on how the promotion of education, and to stimulate research and development efforts, and the help to Gifted people and innovators.

A report of humanitarian development 2002 refer to deficiencies in the construction of Arab development and constraints represented by the lack of freedom, lack of empowerment of women, and lack of knowledge. This report was prepared by the Arab Fund for Economic and Social Development at the United Nations Development Programme, and under the theme "Towards a knowledge society,". The report included multiple concepts elaborated on human development in the Arab countries developments since 2001 for the establishment of knowledge in society, also studied the upbringing of knowledge and the media and the reality of translation, and the links between research and the productive sectors institutions and the role of foreign investment in localization of Technology .

The report introduces culture as an intellectual heritage, the heritage between the construction of knowledge and political recruitment ,then also introduce the religion ,the world, science, , the Arabic language , knowledge society , the crisis in the Arabic language ,promotion of language, and its relationship with thought, also dealt with the protection of quality liberties of freedom of knowledge and intellectual property, and ended up with a strategic vision for the knowledge society, and the report confirmed that the Arab countries have the potential to develop their ability.

The report emphasize that the knowledge is the cornerstone of human development and knowledge-based society is that society which is based mainly on the dissemination of knowledge and production, and employ them efficiently to get to upgrade it once humanity steadily, and with regard to media report indicate that it is the most important mechanisms of dissemination of knowledge.

There is a decrease in the number of newspapers in the Arab countries, the restriction of the freedoms of the press and of expression, violations continue in the field of knowledge production requires the conversion of the wealth of knowledge to knowledge-based capital and the production of new knowledge in the various fields of knowledge, including the natural and social sciences and humanities and the arts, literature and forms of activity.

The transfer and localization of knowledge requires the organization of a catalyst to produce the technology and the production of knowledge through the promotion of scientific research, application of information technology and communication, and the Arabic language be Arab media mass and enhance incentives pattern community that can gain knowledge and employment in the construction of human development, as alternatives to physical possession of values, and finally can develop cognitive

Technology Transfer

This process reflect the countries that have no corner of the technology and include the introduction and import the technology by developed countries to dismantle technology, whether hardware or the modalities and implementation of training programs for skill and buy technologies as well as rights to accept foreign investment.

Technology transfer takes many forms, including the dismantling of machines and tools technically, and then re-manufacturing the technology.

☐ implementation of training programs to take advantage of foreign expertise.

☐ allowing the importer to get patented technology license.

The issue of technology transfer between the different countries of the world is one of the main dilemmas facing humanity and of interest to all parties, whether those exporting technology or importing.

Japan is talking about the need for technological capabilities, while in France a debate about the diminutive French presence in the world Wide Web, and a German, for example, trying to remove cognitive and technological gap between twoparts .

The technology transfer process will either be a new idea or be an infringement of the idea of the present , because it is a move ideas from research laboratories to markets sites, and it is a process of exchange of information between development scientists and users and the global economy is moving more than ever on the economy of knowledge-based technology, and increase economic and social growth dependence on technological level and on the growth of this level.

The weak technological level,its weak growth, two of the most important global economies that led to the increase in the unemployment rate, and lower growth rates issues, and the brain drain, and increasing indebtedness. There is no doubt that the means of technology transfer vary from one country to another according to the degree of development and its ability to absorb and resettlement of technology, as well as according to its relations with other countries and the degree of harmony or belonging to various international conglomerates, and for this purpose different countries to establish specialized scientific centers for the transfer and localization of technology and adopted several techniques to transfer the appropriate technology to meet their needs and according to the following:

☐The development of national centers for the transfer of technology to monitor the movement of the evolution of technology in the world and the impact, and coordination with the productive enterprises and according to the following ;

☐ Preparation technical cadres to implement technological transfer technology in coordination with the various scientific activities of the institutions.

☐ The purchase of equipment and the foundations of its technology contracts are included.

☐ Usage and benefits from the global information databases through information networks.

☐ Investment all opportunities offered by international organizations and bodies involved in technology transfer activities.

☐ The universities follow up the latest developments of updated technology .

☐ Preparation of the technical feasibility studies for various production projects.

☐ Technological talents owners care, and work to refine his talents and Astmthar their inventions.

☐ Creation of technological cities.

The Technology has external sources and internal sources and the technology-transfer either comes from outside or from the inside. The global firms, for example, represent external sources of technology, and the internal sources include research centers and national development. Imad Moustapha, indicates the technology transfer from the report of humanity development and acquisition of resettled technology in Arab countries in 2003.

Furthermore the knowledge and production capacities ,management of technology , ways to increase their effectiveness , rapid and drastic changes in the technology transfer process with the dominance of the currents of globalization and economic groupings mergers big companies and big acceleration in the generation of technologies and intellectual property protection.

The technology transfer become indigenization process and causes pollution and environmental damage, sometimes up to local disaster levels.

There is a new trend to pass statutory environment in developing countries aimed at evaluating the environmental impact of all existing industries and those planned to initiate and created. The technology transfer became today conditional on the adoption of friendly technologies to the environment.

The supplements of Kyoto indicate that donor countries committed to take "all practicable steps to encourage, facilitate and finance technology transfer not detrimental to the environment and to facilitate accessible" in the transfer of technology such as industry companies or agricultural or fishing companies and all investments aimed at energy conservation, production and marketing of energy renewable.

The economic growth based on research and development alone is capable of advancing development rapidly, and are allowed to narrowing the gap between the economies of developing countries and industrialized nations. The technology transfer is to import technology from advanced countries or companies technologically to countries that are importing but the concept of technology transfer in developing countries became interested in the purchase of factories and marketing of its products.

The localization of technology must be able to achieve high control, but the technology development represents the stage following the localization of technology and they are necessary because they allow new invented technologies locally such as products updated by competitive manufacturing. The technology management are also required because it includes the will , the acquisition , use of technology, reverse engineering, local development, dismantling technological package, , appropriate technology for the environment, and ends with the generation of technology that includes research and development, management of the national system of innovation, and patent , intellectual property rights.

There are multiple views and opinions of writers and researchers on the concept of technology transfer, including:

- Automated concept Technology transfer process of production sources and development to where they are used.

• Integrated concept   A series of planned operations of nature (research, extension). The transfer of modern agricultural technologies include:

☐ International Transport for the transfer of modern agricultural technologies, is a transfer of modern agricultural technologies from producing countries (developed) countries to consuming countries used these technology (developing).

☐ Transport (national, local) of modern agricultural technologies.

Technology transfer in developing countries

The experiences of developing countries in technology transfer, including Arab remain stalled as long as it did not achieve the basic objective of the process of successful technology transfer, which can be summarized according to the following requirements conditions:

• A clear and enforceable laws.
• Mechanisms for the implementation of sound contracted investment deals.
• An assessment of the reality of a national technology and the development of technological strategy.
• Implementation of appropriate policies in the field of investment, new inventions and technology transfer.

Antoine Zahlan, shows that all manufacturing and technological acquisition carried out by the Arabs during the past half century did not lead to the any benefit and Arab states may essentially failed to localize technologies imported, and this is normal in the absence of any national policies of Technology. Mohammed Mrayati recommends in his study on innovation and national innovation systems adoption of the followings:

☐ Focusing on technology transfer in the investment in procurement contracts.
☐ Improving the negotiating capacity in technology transfer.
☐ Development of elite in the technology in all Arab countries.
☐ Adoption of a common Arab policies and improving the Arab negotiating position for the transfer of technology.

☐ Unifying the adoption of a certificate of origin for all Arab countries.

☐ Improvement of institutions and the expertise of legislation and technology management mode.

☐ Encouraging the drafting of legislation to finance the acquisition and technology operations.

Human technology transfer

The reasons for the transfer of human technologies and brain drain from the Arab countries out of their borders, including many factors such as "repulsive," factors and "attractive" factors existing or displayed in the host countries, which contribute to the loss of life and experiences and the most prominent factors "driving" or "repellent" are reflected in the following :

☐ Political atmosphere .
☐ The vicinity of the work and the situation of living.
☐ Higher education systems and bureaucracy .
☐ Different technological policies.

The "attractive " factors are reverse the "driving" factors or "repulsive" factors and the most important of these are the following :

☐ Appropriate scientific atmosphere .
☐ Availability of suitable, intellectual, social and political climate.
☐ Adequate level of living for this specialized category and feeling safe and physical well-being and various facilities.

The wide migration for researchers and arabs scientists to the west, referring to the weak interest of their countries in science and research and to do great facilities for the exit in order to open the attractive factors to take them, accounting for 60% of North America of these migratory drain and that for various reasons, including the large number of specialized universities and material incentives advanced attractive climates research and study .

The risk posed by the migration of Iraqi competencies on local development plans, in particular, require solutions to reduce this phenomenon as a prelude to stop them and the best solution is to develop an integrated Iraqi strategy to address this problem with the benefit of UNESCO and the International labour organization , Arab council of economic and social and the Arab labour organization expertise .The strategic plan include the following : -

- ☐ Conduct a comprehensive survey of migratory Iraqi competencies in order to identify the size and location and its fields of competence and working conditions.
- ☐ The formulation of a central Iraqi policy of the workforce on the basis of integration.
- ☐ Establishment of a national program to address the migration of qualified Iraqi situation and centers for research and development and scientific cooperation with international and regional bodies.
- ☐ The organization of conferences for Iraqi expatriates and ask for their help and expertise.

Israel attracts thousands of scientists from the former Soviet Union, engineers, doctors, nurses, artists, teachers, and a large section of them brought with them many of the mysteries of scientific evolution . American journalist "Thomas Friedman " said that (Israel) are classified now in second place after the United States according to the number of new computer-related companies, which spread in the nineties.

Investment in the development of the intellectual capital and constant rehabilitation main factor in the upgrading of technological national policy and counter globalizing hegemony, if any, and requires the development of rehabilitation plans for an integrated and continuous human resources at all levels, including the field of high education ,enrollment and reduce illiteracy , increase the number of educational institutions , increasing the number of staff working in the proportions of Educational field.

Regardless of the estimated cognitive and technological university graduates, the number of employees in research and development is still up 0.01% of the total labor force and the number of specialists in engineering technology who work in research and development less than similar products in many countries in the region.

Education suffers from the low number of students for the technological aspects, which could threaten the economy over the long term. And that this phenomenon can be attributed to the low absorptive capacity of the current economic sectors of the technological education and orientation towards outputs sectors less dependent on technology.

Migration of scientists is due to various reasons, effects some, overlapping in some, and these create psychological climate, related to scientific and incentives conditions. Some of the reasons are social in the homelands, others are materials, may be related to the needs of living.

The phenomenon of brain drain has motives of social, political, and personal nature. The social is characterized by difficulties that faced the developing countries in strengthening the shaken scientific planning in developed countries. One way to keep scientists from migration is to treat the fundamental faults by working to link with national policy, to introduce the idea of scientific planning, providing possibilities for scientific work and atmosphere.

It is an unfortunate fact that the money spent on scientific research and development of the university, in all Arab counties up to 260 million dollars only, while the states of Western Europe during the sixties spent 6 billion dollars per year. United States of America spent 24 billion dollars during the same period. The value of spending on research has increased with the beginning of the eighties to nearly 40 billion dollars. This has encouraged migration of scientists, for example, Iraqis abroad had been attracted to the atmosphere of academic and scientific facilities, methods and the possibility of attending scientific conferences, symposia, magnitude of printed and published by specialized magazines and periodicals.

The phenomenon of brain drain is the most important global problems which recorded at the international level and regional level as stated. Recent studies indicated that the organization for economic cooperation and development which includes 30 industrialized states, that the immigrant enjoyed a degree of education.

It is worth mentioning that the phrase "Brain Drain" derived by British was used to describe the loss of scientist, engineers and doctors and that the UNESCO defines immigration as kind of abnormal types of scientific exchange between the states, as a reverse transfer of technology. The Gulf center for strategic studies in 2004 indicated that western counties had attracted to the west no fewer than 31% of the brain drain for developing countries by about 50% of doctors and 32% of the engineers, and others within the intellectual trends. The west is perceived to the issue of brain drain from standpoint that they reproduce underdevelopment in developing nations.

The risk of brain drain may vary from one state to other, but effects remain similar, in that, the brain depriving in human resources.The "World Organization for Migration" estimates, those developing countries supporting the United States, Europe and South Asia at 500 million dollars annually. On the other hand, the World Bank estimates that one hundred thousand foreigners from industrialized nations are working in Africa at a cost of four billion dollars annually.

As well as that certain social values prevailed in the traditional farming communities would also decline, especially that migration was characterized as external migration of males. Therefore the migration process may lead to partial destruction of the wealth of mankind.Brain drain that began after World War II included developing and developed countries spearheaded by Britain, France, Germany, Sweden, Switzerland and Japan. The United States of America is not included, but limited to become a terminal brain drain, from other countries.

From a historical stand point, this kind of migration was due to Phoenician and Golden ages of Greek, Roman and Arab civilization. In theory, the migration and mobility of scientist across the centuries, from, country to country is considered as one of the features of scientific development. The UNESCO in 1955 considered brain drain as "the impact of migration from the effects of human solidarity".

It is noteworthy, to study the effects of brain drain on the Arab countries in the perspective of a strategy to develop higher education. The competencies move highly qualified group of individuals from one Arab country to another. This brain drain has not received international attention, only at the end of the sixties and seventies following the transmission of certain competencies to the industrialized countries, where moves brain abroad to more advanced society to increase productivity, but at the same time causing a loss for the country of origin.

It is estimated that Egypt had provided about 60% of immigrates to USA, and Iraq has increased its share significantly after the nineties, followed by Syria, Jordon, Palestine. The UNESCO has chosen Egypt from among those most affected by the brain drain, but did not contribute dramatically to solve this problem.

Egypt has used students returning from study abroad in the appropriate places, but Iraq was the first to issue legislation to participate in solving this problem.

Globalization and technology

Globalization is group of economic, scientific ,technological, cultural and political processes penetrated the borders of a single country and led to the convergence parts of the world and the communities differ in the extent of absorption, acceptance or rejection.

Supporter of globalization refers to being a form of simplification of relations between the countries of the world, and those who rejecting globalization and it is a type of control by powerful countries on vulnerable where they stay as a consumer of knowledge and thus the globalization became of negative effects on the economy, education, politics.

And thus facing globalization requires developing countries to return to revitalization of the relationship between education , the economy, development and culture. And with this, the globalization has positive effects on education as popping a growing number of universities linked to a network of Internet offers its services globally, as well as the emergence of university alliances across the border.

There are more than 200 gathered for a global university alliance such as Beijing University , Seoul National University and the University of Tokyo. The pressures imposed by globalization makes it necessary to allocate sufficient resources to the higher education sector, and must be reform of the sector at the enterprise level . The Higher Education with the help of developing countries, and benefit from globalization through technological developments, can to be its tool essential because it helps developing countries to reap the benefits of globalization, and finally connect the globalization of higher education provides opportunities to improve living standards. India, for example, benefited from globalization by building computer industry.

Some believe that universities and higher education international perception of nature and global the trends, not globalization trends, where the world is different from globalization, higher education is in the context of globalization is seen as a commercial commodity, governed by market forces, if globalization has positive aspects from the perspective of increasing access to higher education opportunities, and reduce the the knowledge gap in developing countries, it has at the same time negative aspects threaten universities States as well as it has affected globalization extensively in universities as a result of the policies of globalization.

The support of the public sector in the free market economies, led to universities managed commercially very clear globalization concepts, exciting always controversial, they are economic, cultural, media, and scientific technological and cultural globalization without political globalization imposes obligatory intimidation . Technological globalization of political principles, economic, social, cultural, and it may represent a significant development in the communities with the degree of penetration and it must be played criteria and indicators of economic strength and social stability and progress in all areas of life and this requires repair work rules in developing countries.

The phenomenon of the illness requires the national entities thinkers discuss these perceptions with the best ways and means, departing from autoimmune conditions and privacy in order to create a national vision is able to absorb. Representing the Japanese American political framework of the globalization of ideas, has mentioned in the book ((history and human)) for the policy statement of the new world ,globalization its philosophical theory in economics, governance and life and domination education and technology background.

The study of technology is one of the main areas of learning and teaching in the educational curricula brought forward so it should be adapted to the school entrance, which can extend the learners' ability and competence to enter into the field of technology beginning from kindergarten to the end of high school,

Many of the evidence in the educational understanding require developments of new technological tools and a variety of help, if one of the education goals of the production of knowledge, the purpose of the technology is to solve scientific problems through innovation means and systems, procedures and environments that show how to improve the people for their lives, so the technology Modern rely on farming as they often require knowledge that could be made available through scientific research.

Education is technological is not in isolation from other substances, especially science as the concepts and skills technological rely to learn and develop the deep understanding of the facts, concepts, principles, laws, scientific theories, and the theoretical study of these concepts and principles without the benefit of them in technological applications, makes it possible to take advantage of them is It is not possible, so the technology is considered the practical side of science.

In light of this occurred in the teaching of science many changes in light of technological changes did not teaching science is limited at this time to train the learners to solve nonlinear problems in which the learners follow a specific path under the guidance of the teacher until they reach the result or the expected solution, and thus do not have the opportunity to creativity and innovation, but with the emergence of the concept of technological education healthiest teaching science seeks to achieve the objectives more closely related to the environment and the demands of society and limit exceeded teaching separate scientific knowledge parts, to focus on the implementation of many of the scientific and technological activities, including gaining the learner the ability to cope with real-life problems efficiently and effectively and enable it to dealing rationally and logically with the available environmental resources in order to achieve its requirements and the requirements of society, at the same time achieves a degree of autonomy in resource investment and employment in order to achieve some sort of self-esteem and this is what takes the teaching of science in the Arab world in general, where you must get rid of the traditional methods, whether in the design and drafting of content, or in the preparation of scientific experiments that learners know the results before they are implemented, or educational activities that do not challenge their mental abilities that are evident when dealing with the emerging modern systems technology, so there must be a more comprehensive and depth in science education and that by providing an opportunity for learners in all levels of education to see the scientific and economic value of the life of their learning of these concepts and theories and scientific laws

The ultimate goal of scientific breeding technological is to prepare human resources to deal with the machinery and equipment production large, they are focused on educating young people from an early age on the technological sense, creativity and innovation, by relying on what provided the educational environment has the potential and learning of facts and concepts and scientific principles and theories and laws, and here come the learner psychological, educational and educational balance while finding pleasure in learning and interest back on him and his community of this education, and this is the real difference between the educational systems in developed societies, which in its philosophy and objectives seek to achieve this balance the learner through reduced the gap between the scientific and technological aspects and even get rid of that gap, while the case in educational institutions in the Arab countries, we still find there is a large gap between the scientific and academic side and the side technological, and the reasons may be related to non-availability of the awareness of those in charge of public education systems and specialists to prepare and formulation of science curricula importance technological enlightenment of learners and the need for technological capacity and skills development of learners, either for fear of failing to strengthen science education or technological education mechanisms, there may be reasons may be unknown to the author.

Science education technological interested in developing an individual's personality / learner to suit the spirit of the times being experienced on the one hand, and the needs and the needs of society on the other, which is an integral part of general education in part, they are integrated with other breeding patterns to pour all of us in this direction, scientific technological Education is seeking to achieve its objectives as well as to contribute to the achievement of other educational goals, they do not operate in isolation from other educational areas, "or various educational topics, but must go through all of them in one consistent and in harmony is no contradiction is not opposed to them." And which require the use of so-called mental maps nature standard questionnaire to resolve these issues.

"Technology as a science applied" reflected on education and philosophy and its system, as well as traditional topics such as industrial arts or crafts, where we find that factors of technology had infiltrated the general scientific education and special education technology and culture .

The Arab Organization for Education, Culture and Science, and ordinances prevailing concepts which is influenced by the members of the community in various categories of values. A report of Glenn Seaborg entitled - Education for the era of science, the vital role of scientific culture in contemporary societies, by stating that the citizens in a democratic society today should understand the science in order to have a wide participation and clever in many national decisions therefore scientific culture have of diversity and plurality at different levels either.

These deployments scientific and technical culture, it may three dimensions:
- Dissemination of scientific content.
- Definition of knowledge.
- clarification of construction-related cognitive factors.

From this we can say that the main objective of the deployment of technological culture, is a support community to enter into a fertile ground for the production of scientists and skills and the implementation of the following:
- Scientific transparency.
- Absorbance of technology platforms.
- Creation of sympathetic environment with technology.
- Qualitative development of individual thinkingof .
- Achievement of technological security and production of scientific knowledge.
- The establishment of innovative mechanisms to employ knowledge.

The scientific culture that it may contribute to the organization of thought through a variety of criteria including:
- Movement from the unilateral opinion.
- Cognitive humility.
- Scientific integrity.
- Scientific rigor.
- Ethics and quality of work.
- Generation of motivation of intrinsic knowledge to continue.
- Honoring scientists and innovators.
- Rooting of democratic values.

Dissemination of technological culture requires a package of measures including:
- Creation of scientific atmosphere .
- Support of strategic planning.
- The development of a knowledge society.
- Fill the gap between the scientific community and the public.
- The study of the ethical and social dimensions of technology by discussion.
- Simplification the technology.

- Establishment of science clubs and museums.
- Encouraging of scientific competitions.
- The provision of simplified scientific books.
- Rehabilitation of scientists in vital areas.

International experiences in technology

Since human history until the advent of monotheistic religions believe that there is huge potential carries with it a fantastic solutions and piercing the disposal of the problems being experienced and the technology has been around since that man was found and the last of the creatures and since found in this universe which is in conflict with nature is driven by the instinct Love to stay aware of scientific or mental of the need to improve his standard of living and stability and the Secretariat while the bees use in their daily lives, such as building cell and ants drag their diet and that technology does not take its real role, but with the only man because he who has the ability of invention and creativity Man adapts to everything man has clearly been outgunned in the field of technology lies in the coordination between mind and hand tools to prevent material and the possibility to control the time and place for it represents the collective memory in which to accumulate

knowledge and human experience.

Technology dates back to about half a million years old when man discovered fire was invented and demonstrated a shining talent unique in exploiting the opportunities that stand out in front of him and such a technological saga that line had rights to control the nature peaks and across all human progress and the metal industry to develop a social life and a source of light to protect the rights of animals and leading their tools technological like fire and ax made of flint, wood, needle made of bone and twenty thousand years man-made set of tools that formed the first and Ayers communities their tools technological fire ax and Stitch considered technological revolution and before eight thousand years since man began planting fields and the domestication of animals and form caucuses clear that the evolution of technology has been associated closely the emergence of agriculture, which focused on the banks of the Asian and African rivers in Mesopotamia, Egypt, India, China, before about 1200- 4000 years the evolution of T. technology turn and opened to these civilizations horizons and new limits and managed these civilizations and especially the civilization of Mesopotamia and the Nile Valley to be a pioneer in the creation of technology relating to mining activity and vehicles animate tires and metal spirals pottery rapid plow metal and brick industry and the use of papyrus and melting glass.

The Mesopotamian cradle of technology where indicated investigations of archaeological on the agenda of architecture and systems of irrigation, water, roads and planning cities. The laid the emergence of the New Stone Age in 2500 for BC foundations technological major civilizations that followed and which extends to the civilization of today and planted a lot of scientific seeds and awareness of human and major civilizations Subsequent and the Arabs and the Europeans who carried him to a new height.

The saw New Stone Age one of the main achievements of technological in all the pictures, the discovery of agriculture and the domestication and breeding some of the animals and this development constitutes a historic shift in the economy and the fate of man and mankind emerged improvement in the technological tools that require the nature of this activity polished stone tools and sickle needle and saw the house, pottery, plow industry has emerged has approached the New Stone Age to an end real crisis as a crisis of food entered and became farmland relatively rare due to the increase of the population everywhere and turned the ax of harvesting human beings, wars and bronze creates Sbv and became a human enemy of man and began a new phase of history

And Iraqis invented the clay figures to record cuneiform characters as it was for the Iraqis and other great achievements and invention system sixtieth time, and in Mesopotamia Engineers Babylonians reach a basic arithmetic operations and algebraic, and account different surface areas and volumes. Also managed the construction of buildings, bridges, roads and pavement, before more than three thousand years BC.

In the Nile Valley and the ancient Egyptians managed to achieve engineering and technological achievements been building the pyramids, which are excellent examples of the skill and precision engineering calculations and good execution. Egyptian Engineers also completed important works in irrigation and land reclamation. The ancient Egyptians excelled in mathematics as indicated by the patches of papyrus dating back to about 1500 years BC, which refers to the knowledge of the ancient Egyptians triangle and calculate areas and volumes.

With the fall of the Romanian Empire civilization again moved to the Arab countries, where have achieved important scientific achievements, including the development of chemistry and optics science and the creation of paper, sugar, soap and perfume plants that have become part of the culture of that era and the Golden Age of Technology Atd Arabs took leaning towards decline and decadence in the same direction as the Arab civilization entire internal Valtvkk The civil wars and attacks the Tatars and Mongols and Turks are all Crusaders contributed to grab civilization of the Arabs and the deterioration of the Arab-Islamic civilization moved science and knowledge back to Europe and moved torch of civilization from the hands of the Arabs to European hands by what is known as an era of renaissance in the objective conditions, especially in the field of technology and seemed trend towards maximizing the mind in humans and its ability to innovate and is worth mentioning that the Greeks have benefited a lot of close contact Eastern civilizations peace or war, and organized scientific thinking theoretically and imaginative and abstract and a circular began with the Greeks at the level of theoretical thinking not been matched by comparable level practical applications of science that invented any on the level of technology has created a gap enormous between these two achievements .

Arabs and scientists combine theoretical and applied Kalpouselh Kjabarbn Hayyan in chemistry and Canadian in optics and Razi in medicine, Ibn al-Haytham in optics, physics and the Peronist in physics and Ibn Sina in medicine and others where these scientists have developed all these scientists and other scientific methodology does not differ in something on the methodology followed by Europe in the renaissance of medieval dark and disasters brought by the Middle Ages to the European human that made him moving more and more towards rationality in dealing with everyday problems and contributed to the development of irrigation methods and in the water and wind energy adopted by Europe at a later stage use Arabs also creative in technology ceramics and stained-glass method of refining sugar and the invention of the catapult and mobile towers. Able political and religious crises and the spread of diseases that cause significant damage to the population of Europe at the end of the fourteenth and fifteenth century and remained throughee to the surrounding environmenton. And heat and a calculator and pump air and there are different factors that contributed to the development of the seventeenth century, including the dialectical interaction between people's perception of the religion and the restrictions that were imposed by the Church on the economic aspects and the private capitalists who relied on the machine, which was allied to them, as well as the evolution of the technology was until the middle of the eighth century Ten due to the discovery of craftsmen, technicians and skilled workers so taking illiterate workers carting face increasing difficulties in following up the developments have Black and contributed two examples of early goals of the scientists were able to explain the theoretical basis for inventions reached experimental ways and in the nineteenth century has become a science and technology is susceptible to separation

.

The evolution of technology with during previous eras go at a slow pace but there is a huge leaps occurred after World War II require deep analysis to keep pace with economic and social their effects and the history of technology shows that the gradient in this area was more than any other area human nature and technology development aid it achieved gradually and curvy. First Man was realistic and pragmatic raw materials available to him making tools that increase his productivity and power use stones, bones and wood to be the first Mafi repertoire of technology.

Developing and Arab countries have failed them in the traditional manufacturing policies did not lead to the localization of the technology transfer and did not lead to push forward economic and social development, either new trends developing countries on promoting foreign direct investments require that the planned properly and create a knowledge flow environment contribute to strengthening the structure with research and development, which allows the transfer of the Arab countries of resettlement phase to generating effective participation in the global technological and structural phase.

The technology transfer operations and development by resorting to foreign direct investment will continue to be susceptible to Problems in the transfer of Arab rentier economy to a value-added economy, but to a knowledge-based economy.

Economy and Technology

As we mentioned earlier in one of the definitions of technology " Stock of available knowledge of a society at a given moment in the field of industrial arts and social organization embodied in goods and styles productivity and management at individuals, institutions and the state", the change or technological advances is to improve the level and quality of technology available, production of discovery methods new goods and previously unknown geometric designs and innovative. Technological progress of scientists "Technological progress is one of the most important factors responsible for the economic growth, if not the most important at all. The increase in per capita income in addition to elemental labor and capital, also dating back to the technology, especially technological progress.

In addition, factors of economies represented by
- Management improvements
- Improvments of the level of education and technology

Historically, the technological progress in the United States of America during the fifties contributed about 90% of the increase in economic growth, and is worth mentioning that technological advances and improve the level of education and technology and the other, called factors intangible and indirect a larger role in the steady rise in the standard of living in the US United. that the world is moving towards a knowledge-based economy and explain this trend to modern economic theories, including the new growth theory. This theory says that sustainable growth (not growth for a short period) directly depends on three factors are technological level, technological growth and the percentage of savings. And traditional factors in a growth of capital and manpower intervened in the growth equation indirectly through technological level and technological growth. This is very important changes in the global economy is expressed in another way is that the high value added comes from the technological level and of technological development of the State and not only from the capital investment and the number of labor. Economic growth certainly depends on researchers who discover and invent and to those who invest these inventions and creations. In addition, the good implementation of these innovations depends on skilled workers who can deal with the modern means of production. That many countries no longer rely solely on the events of industrial zones and free zones and the like because this mechanism no longer provides an acceptable value. But it began to these countries since the eighties adoption of new patterns are technological areas and cities of science and technology and areas of knowledge. The objective basis for these patterns is to make the most of new ideas and creative and issued by the research centers and universities and support these ideas and bring them to the production lines assigned to his service. The link between research and the one hand and industrial activities and service on the other hand is the nucleus of a statement of these areas or cities Grandpa

As we mentioned earlier in one of the definitions of technology "knowledge available stock of a society at a given moment in the field of industrial arts and social organization embodied in goods and styles productivity and management at individuals, institutions and the state, the change or technological advances is to improve the level and quality of technology available, Kokchav production methods new goods and previously unknown geometric designs and innovative. Technological progress when scientists "Technological progress is one of the most important factors responsible for the economic growth, if not the most important at all. The increase in per capita income in addition to elemental labor and capital, also dating back to the technology, especially technological progress.

Informational technology

Iraqi universities and in very limited building an information network through which the provision of online services is worth noting that the percentage of the number of students to the number of devices available is too high without a general framework for plans chaos in the development of these plans, which are most often repeat to each and in the light of these random to be put the general framework which defines the requirements and terms of reference to be developed

Suffer universities and colleges from an acute shortage of specialists number of related science and Computer Engineering as well as that software which is facing many difficulties and challenges that limit the economic feasibility and benefits entry of computers in the field of education puts the planners of higher education in Iraq on the eve of social and economic changes as well as changes industry education that has to be absorbed and dealt with to create educational systems can keep up with her renewal and development of Internet networks extensively.

Appeared in the last quarter of the twentieth century technology "continuing education" and did not features identified and clarified its properties, but since (1960) when the Second International UNESCO Conference on Adult Education held in Montreal, Canada, the city, and it put the first touches, as it was decided that it is no longer enough to spend specific individual years in school education. In the year (1964) and approved by 119 countries in UNESCO's General Conference on the recommendation of the effect that different forms of education outside the school and adult learning should be considered an integral part of the education system to have the opportunity for males and females to continue in lifelong education and still patterns of Continuing Education in Iraqi universities suffer from prejudice part and continued weakness in the performance in spite of the long period practiced.

Open universities receiving students from different educational levels to provide them with types of studies required by the developments of the individual and society, and suffer from the weakness of Iraqi universities to accept this kind of education.

Data banks

Information banks are considered an outlet conventional storage problem and can retrieve information stored in these centers using one of the elements Description such as document number, author, title search, and aimed information banks to link the centers and scientific institutions and universities by the light in front doors of knowledge and information they gather their international experiences. In Iraq, there are no techniques and even the terms of this education.

No way in front of the Iraqi universities only option for the development of technological change, and specifically technology knowledge which is considered the most important factors helping to produce knowledge, which offers technological logic and then access to the production of more advanced technology.

Reinforced developments in the field of technology and communications from taking open university education pattern, due to its reliance on knowledge and information technologies. Given the rapid development in the communities and the transition to a knowledge society, and knowledge-based society will see a significant expansion in taking this type of university education, with the justification that we have mentioned previously, including: -

- integrated use of other modes of technology
- submission services to individuals of all ages.
- beating barrier place.

Technology knowledge and its relationship with the Iraqi university

There is a clear relationship between knowledge and technology knowledge society they are two sides of the same coin, The role of the knowledge society is clear and formed his deportation as a result of stunning advances in technology applications, where communications have developed a stunning view of the evolution of technological turning technology to revolutionize represent one of the important innovations done by the man at the end of the twentieth century and the beginnings atheist and the twentieth century, then expanded this session in the necessary areas to humans, including:

- chemistry and medicine
- Life Engineering
- Other Sciences

And contributed to a substantial change in the lifestyle of modern man has become the mind and the thought is the basis of profit and investment, and therefore there is sufficient justification emphasizes the importance of and the need to take Iraqi universities this type of education. To meet the challenges of the learning community, must pay attention information and build databases of knowledge and networks of modern communications and the integration of technology in teaching and learning and research operations, and that the success of universities and turn them into learning societies depends on how much attention in the faculty development professionally and adoption of the principle of participation and planning and activation and the use of modern technology and transform the classroom to the active and effective learning environments established over the past few decades, a lot of technological research institutions in Iraq as the body of Atomic Energy and the Public Authority for Agricultural Research and varied these centers capabilities and potential, highlighting acceptable to each other and the role of the absent for some other effective as a result of administrative, financial and legal .

Obstacles that there are some technological institutions duplication of tasks form which required re-arrangement and structure of institutions essential to the next phase to maximize the added value derived from such institutions as the establishment of new components is very necessary scientific and technological such as cities and incubators specialized Public Authority for biotechnology and the General Authority for Remote Sensing of the General Authority for Information Technology and the National Center for Energy Research and other . Justifications and motives of drawing national policy of Technology.

Malaysian experience spread in various cities, villages and mobile libraries, also offers great facilities for students and talented, with great support of scientific research institutions and universities, which Malaysia has moved from a primitive agricultural country into an advanced country occupies ninth place globally among countries exporting high technology, and that an annual growth rate approximately eight percent of various sects and ethnicities live in Malaysia population familiarity, cooperation and sharing competencies in a friendly way, his background has given him motivation technological interesting turn of Central . other prospective future Malaysian program "establish a full community of values ,.

Technological incubator

The global economy as mentioned is moving towards an economy based on knowledge and explain this trend to modern economic theories, including the new growth theory, she says this theory that sustainable growth (not growth for a short period), directly depends on three factors: the technological level and technological growth and the percentage of savings . The traditional factors in a growth of capital and manpower intervened in the growth equation indirectly. And it expresses these very important changes in the global economy that the high value added comes from the technological level and technological growth of the state and not only from the capital investment and the number of labor. Economic growth certainly depends on researchers who discover and invent and who invest these inventions and creations. This is in addition to the good implementation of these innovations depends on skilled workers who can deal with the modern means of production. That many of the countries are no longer just depend on creation of industrial zones and free zones and the like, because this mechanism no longer provides added value acceptable, but these countries since the eighties began to adopt new patterns are technological areas and cities of science and technology and areas of knowledge. The main objective of these patterns is to make the most of new ideas and creative issued by research centers and universities to benefit.

Chapter Two

General Aspects of Technology

contents
- **Preface**
- **Technological planning**
- **Technological reforms**
- **Foundations of Technology**
- **Novel challenges for Technology**
- **Global experiences in Technology**

Preface

Many of the definitions of technology lack methodology to the intellectual aspects, and thus require that the definition is based on objective grounds including the philosophy and trends of humanitarian, political and scientific developments. However, the definitions are consistent with this perception, including that of technology set-ups help the community to clarify the goals and heritage on an ongoing basis..

The public perception of the definitions of the former supposed to be matched more precisely the role of technology . Modern technology reflects the effort sought by the community group activities, multipurpose development of a new generation or developing a new generation. The stages include the technology of modern community and focuses on the second experience by the community.

Technology is different from one society to another as well as in the same society in terms of the political, social, economic, cultural and scientific aspects. The technology focuses on the scientific example. Some thinkers suggest that technology is not transmitted from generation to generation because it is inherently adaptable, more complex.

History of education technology varies according to the doctrine of technology . Western technology , leaving the individual free, socialist theories developed technological policies by the state. It should be noted there are different tendencies prevailed since the nineteenth century and the mid-twentieth century, including the psychological tendency and the tendency of scientific and social trend. Concerning the relationship of technology to development that represents a set of processes and procedures at the theoretical and practical lead the entire society to change as well as a dynamic behave in several ways.

There are different types of technology based on the different ideas that divided them such as national integration, balance and continuity, government civil and parallelism, distance and proximity, innovation and comparison and information and economic development.

Technology has many goals for the purpose of exercise of upgrading the level of individuals and society, mature through plans of action and interaction between individuals and society. There are particular targets of one community because of the special features for the specific community and general objectives of the other reflecting the variables nature of society, as well as that some think that the goal of education is supposed to be final and not focused on the concerns of learners and emerge from the community and experiences resulting from the interaction of its personnel..

Technology goals are classified according to the human evolutionary stage targets of primitive societies and the desired qualities and purposes of religious knowledge, democracy and the life, scientific and national issues. There are behavioral objectives have been defined by educators as the desired changes in the behavior of the learner.

Goals of the educational technological system are linked to social and political system of any society that can face social and varied economic conditions that should seek to achieve. The channels of the technological goals of the any system are determined according to the quality of the social system and philosophy and concepts.

The technological system is also composed of a set of secondary sub-systems, including governance, funding and scientists training system and the system of examinations, technological philosophy, , and in general is a collection of elements that governed relations technology. Effective technology system is an open system capable of change and renewal of the amendment so that the input and output permanent improved mobility.

The features and qualities to any public technology system are determined by the existence of technological institutions that depend on the growth of life and maturity in different capacities and in the starting conditions and the capacity of the technological system.

Technological administration is always linked to the state system, and differs from one social system to another system and as a result of centralized and decentralized.

The teacher has an important role in the field of technology and the need constantly to large numbers of scientists, so it is important to care for the type of researcher and raise the level of concern through training and improving the economic and social culture, taking into account the public, private and professional.

Of the most important modes of technology in which it plays the most important role in the upbringing of the scientist, as well as the school which is considered one of the valuable institutions on the global civilization. The non-specialized technological institutions include the media, popular organizations, institutions organizations and the social and professional organizations of a formula.

The technological evaluation determines the progress made by the researcher towards the achievement of the goals of technology. Technological measurement is a mean of evaluation that includes an array of stimuli prepared to be measured. The areas of technological assessment are manifold, including evaluating the work of the scientist, working in the fields of technology evaluation, clerical administrative, plans and technological policy, and others.

Technological planning

Overall, planning is a means of resource management and coordinating mechanism or actions taken by the actors or units at different levels of decision- making (government, enterprise management).Planning can be divided into different types according to a number of criteria and scientific phenomenon that has a special significance related to humans, in preparation and development than the other types in many ways, including its nature, origins, methods, comprehensiveness and effectiveness.

Accordingly the technological planning is a meaningful activity and behavior of diverse growth-oriented not rely only on a personal level or at the school level is organized and sustained to achieve future goals through appropriate means. Then it is supposed that the functions and requirements of planning, includes the objective and the value of the implementation, the goal and assess of the results, the adoption of activity of the most effective and human costs and time and financial resources.

Historical changes started to appear on technological planning in general terms while the Parthians were busy 2500 years ago as philosophers scientist and religious people [philosophers utopia (Platonism "Platonism") about life], as well as the Scots "Scottish" who planned in the sixteenth century for the material well. The contemporary philosophers have planned many great experiences, such as those obtained for the experiment of communism and capitalism and the experiences of National Socialism.

Technological reform

The programs of technological reform is prior to economic reform, promote private sector, create real opportunities and to bring technology and increased exports. The state administration of economic activity should be in the narrower borders and must therefore reconsider the role and size of the public sector.

Technological reform programs such as the partnership have complex images including the level of the technological institution by conducting advanced research in the community for the benefit of intra-service institutions, productivity, and others as well as to provide a variety of consultations. The partnership that include the technological institution and the community represent an essential input for the financing of the technological institution to community and contributing to subsequent improvement in the level of their performance as raise the scientific stature, and solve the many problems and provide appropriate alternatives to it.

Technological reform is complex with many subjects, multiple, complex and difficult channels, and therefore the process invites us to look for a lot of reasons for the adoption of planning to reform the technology.

The planning is the first in all processes aimed at integrated human development requires from working staff in the field of technological planning, knowledge
Among the methods of the areas ,modernization in both the curriculum and administration of technological and professional growth, , methods and sources of funding, evaluation methods, , qualitative and quantitative expansion in technological institutions and thus transferred to the development of the finest in theory and practice and through research on the technological philosophy and in accordance with those reform process of education is a systematic process .

Technological reform, including the promise of change and development and renovation of technological institutions and the types that include the renewal and has multiple mechanisms, the transfer and adoption of scientific knowledge, new technology (to strengthen and enrich the programs of teacher preparation courses, training programs on the use of the Internet).The most important sources of technological reform are the motives and rationale for having the institutions to carry out its functions,an imbalance in the movement of the interaction of inputs and low quality output

Moreover, there are general conditions that require their presence in carrying out reforms, including e technological reform to happen after scientific studies derived from the culture of the community and providing the necessary funding in the assessed budgets, original aspects of the culture of the community of religious and spiritual values, language, and history.

The UNESCO has chosen cases involving technology reform such as evaluation of the needs and professional unemployment. The contribution of technology to the development in the Arab countries depends on an effective technology that requires reform. The reform program for included technology such as: (partnership) is based on the impact, vulnerability, and joint mutual work between two or more types of technology related to the same goals, plans, standards and values.

The technology funding in developed countries does not support on one source in different sectors of society, while funding supports technology in developing countries in particular government as a principal source of funding, but that these governments have experienced economic crises in a row, making it unable to meet to secure the necessary funding to technological institutions in order to achieve its goals in the community.

The government support is known globally as the basis for scientific achievements and complement to the role of the private sector. The industrialized countries allocate funds from its annual budget for research and development to support programs and projects that have social and economic projects with the high cost which includes high risk that did not contribute to the Iraqi government. But that the United States of America, for example contributed to budget support scientific research for the year 2004 (122.7) billion dollars, half of it for defense research while research projects of life sciences and medicine has been estimated by 27.7 billion dollars.

The National Science foundation, which means basic science research has reached 5.5 billion dollars, then distributed the remaining funds on research in the institutions of NASA's Space Science and the Ministry of Energy and the Ministry of Agriculture and others. These systems has become a reference on the subject of free technological studies which was not having clear visions but a clear policy at the international level in the sixties and since then began a gradual shift on the subject.

Foundations of technology

It is intended that basics of philosophical application of theory in the field of technology for the purpose of clarifying the process and to identify its features, which is also being organized as intellectual activity, depends on the philosophy as a tool to organize and coordinate the process.

The concept of technology differ according to the kind of philosophy that deals with it, indicates the formulation of the human being announced and well-being, and Plato believed that is the consistency between the self and body, while Abu Hamid Al- Ghazali believed is a priority for humanitarian and spiritual atmosphere.

John Dewey, one of the philosophers, who pointed the principles underlying the concept of technology , including that education is a small community that comes to life and technology continues forever, and the curriculum must be accompanied by life and mission of to prepare the individual for life.

The subject of philosophy is knowledge of the natural facts (relative) standard and the facts (values and ethics) and the philosophy is considered as perceived by Dewey. For example, the general theory of technology is characterized by a turning humanity of human beings to the intellectual and others believe the philosophy Celebrities are the leaders of educational thought Ksagrat, Confucius, Plato, Jean-Jacques Rousseau, which reflect about the conditions of philosophy of technology and work to evaluate and critique the process.

As the fields of philosophy and issues of our time metaphysics (metaphysics), or the divine science, knowledge, values and philosophy of technology is to apply the positivist approach in the field of human experience which we call technology. The community is composed of two layers, one of them think and other works in other words, the first class is linked to the technological framework for the purpose of access to knowledge, which requires the mind, and thus knowledge is a real constant and does not change. The philosophy believes in and basic principles, based on the ideas of absolute belief in the existence of independent real in a perfect world.

According to this philosophy of Aristotle (384 BC - 322 BC), that there is only one world, that is the real world, characterized by the firm principles and the care of the senses. It is more important than a focus on imagination and consistent with the ideal of the virtues that is fixed. These realistic narrators follows stability approach and this philosophy depends on the continuous attempts to discover the universe and the world and work to understand the existing laws that include all the facts within a stable and consistent world.

Pragmatic philosophy plays key role in the development of teaching methods and improving the traditional method and manner of trial and error. The school of John Dewey is based on the philosophy of pragmatism played historical intellectual revolution against traditional schools, which focused on information and not believes in the values of morality, but according to the relative renewed convictions of society.

The education in the pragmatic philosophy is not a process of transmitting knowledge to students for knowledge, but help them to meet the needs of the social environment and able to raise the forces required by the student, including social attitudes. Perceptions of the pragmatic philosophy depends on building curriculum writing arithmetic means rather than targets, and approach interests in the facts relating to the nature of the child. Pragmatic philosophy stresses on the development of technology, natural sciences, while the humanities and languages is of secondary importance.

Novel challenges for technology

New challenges are witnessed by the technological revolution and the production of knowledge as well as the economic challenges and globalization. Technological revolution helped to create a new reality in the fields of science, knowledge, information and communications, resulted in forcing , reconsider their curricula and economic content.

The major global challenges have merged, clearly defined distinguished and experienced on technological institutions. Policies, led to the introduction of significant shifts in scientific developments, social and economic. These challenges are as follows:
- Globalization
- Knowledge
- Quality and Evaluation in technology
- Transfer of technology Brain Drain

The phenomenon of brain drain has motives of social, political, and personal nature. The social is characterized by difficulties that faced the developing countries in strengthening the shaken scientific planning in developed countries. One way to keep scientists from migration is to treat the fundamental faults by working to link with national policy, to introduce the idea of scientific planning, providing possibilities for scientific work and atmosphere.

The phenomenon of brain drain is the most important global problems which recorded at the international level and regional level as stated. It is worth mentioning that the phrase "Brain Drain" derived by British was used to describe the loss of scientist, engineers and doctors and that UNESCO defines immigration as kind of abnormal types of scientific exchange between the states, as a reverse transfer of technology.

The "World Organization for Migration" estimates those developing countries supporting the United States, Europe and South Asia at 500 million dollars annually. On the other hand, the World Bank estimates that one hundred thousand foreigners from industrial nations are working in Africa at a cost of four billion dollars annually.

Brain drain that began after World War II included developing and developed countries spearheaded by Britain, France, Germany, Sweden, Switzerland and Japan. The United States of America is not included, but limited to become a terminal brain drain, from other countries. From a historical stand point, this kind of migration was due to Phoenician and Golden ages of Greek, Roman and Arab civilization.

Globalization is a group of political, cultural and economical operations that penetrate the borders of a single country, leading to the convergence of parts of the world, and leading to differing views of intellectuals on the concept and objectives, and societies differ on the extent of assimilation, acceptance or rejection. A proponent of globalization refers to a form of streamlining relations between the nations of the world. As opposed to globalization, it is believed as a balance between actors, the strongest is at the level of global capital that dominates large companies, and also it rejects the view that globalization is a kind of domination by powerful countries on the vulnerable, where consumers have been the last of knowledge and thus globalization have negative impacts on economy, and politics.

In the face of globalization it requires developing countries to revert, then to revitalize the relationship between education and economy, development, culture. Then technology and economics of future generations are the basis for developmental spirits of the knowledge society and culture.

Globalization is an economic process; some believe that its goal is to make the world a single market. It is transnational companies without a national capital loosely defined administered globally; depend on models of cultural phenomenon that blurs the cultural and linguistic distinctiveness. Some believe it threatens the world's languages of extinction. It has entered under the term globalization in the early eighties the publication of a book entitled John ten major trends are direction of change in out lives, since this phenomenon has become a popular cause of hypocrisy and debate and then the definition of specific and agreed-upon is not easy.

Accordingly, there is lack of agreement on the definition of globalization because it is a multidimensional phenomenon aspect and in that there are factors that contribute to the emergence of globalization. They include the development occurred in the field of information and communication technology, the emergence of the internet and increasing trends with the composition of regional blocs and the major entities (the European Union, the World Trade Organization), as well as the emergence of global quality standards and innovative problems such as: unemployment and environmental pollution.

Globalization has led to the extension to every aspect and all aspects of life, to affect each other and can achieve the goals of the organization of globalization the role of communication and communication technology (satellite channels, giant corporations, media, and internet).

Globalizing operations include political, cultural and economic penetrate the borders of one country, led to the difference of opinions among scholars about the concept and goals of globalization that refers to being a form of simplification of the relations between the science. The view that globalization is a kind of control by powerful states on the vulnerable, where the years have negative effects on the economy, technology and politics.

Globalization has also expended in a restructuring of global capital and guidance to multinational corporations of manufacturing industries, especially those that aggregate operations are characterized by highly labor-intensive. High foreign trade flows of goods, services, the movement of capital and information and personnel between the international border of the countries as well as higher levels of integration of markets between countries.

From a strategic point for the preparation of human resources for the knowledge economy require the identification of competencies (knowledge, skills and attitudes) to various levels. Man is the result and purpose of development and the instrument and the decisive factor in the development.

The human capital includes skills, talents, abilities and knowledge of the individual elements of the evolution of the individual contribution in the production of goods and services. The investment is to postpone current consumption in order to achieve a return in the future, either about what motivates human capital accumulation is the likelihood of achieving the ability of higher income and this is the financial return to justify investment in human capital, whether by individuals, employers or society at large. The investment in human capital, formal education and job training, note that there was a problematic in education that can distinguish it from the investment side and indicates economics of education to adopt the following equation refer to age, technology, income

$Y1 = P + BS1 + X1S + Mi$
Y= incomes or wage
S= study
X= set of variables that affect income
M= disturbance term

In order to calculate rates of return to investment in education, we must compare the relative cost income at every level of technology . It is required that technology should be change to meet the challenges of the economy in the information age and if we want students to become intelligent users of technology and information, they also have to accept how to be creative and innovators, where they must solve problems able and understand how to analyze date and reach conclusions smart and know how to use new technologies and excellence and equality.

The investment reflects in learning to the scientific research, development budget, the proportion of scientific research and development to total gross national production. The investment in scientific research is called which dates back by huge returns commensurate with the value of spending the knowledge economy. The world's investment leads to increase in research and development for the period 1990-2000 from 410 to 755 billion dollars. It is noted that developing countries including Arab and African countries located in the lowest density of spending on research and development on the other hand the budget of the United States amounted to U.S. government support of scientific research to 122.7 billion dollars for the year 2004 in the areas of the National Health and National Science foundation and the foundation for NASA space science, energy, agriculture. The private sector has contributed to the spending jumped at the drug companies from $ 1.3 billion in 1977 to 32 billion dollars in 2003.

Global Experiences in Technology

. The development of technology in Japan was carried out through the many stages, the antiquity and the stage of European culture and friction, . Historically the evolution of contemporary modern technology in Japan occurred over during three periods:

• The first period (1872-1939) witnessed the first renaissance in the nineteenth century. Japan quoted in this period the French system . Japan was based on the centralization of the nationalism and the emperor, turned into a kind of militaristic that was able to defeat the Russian army and the Chinese military forces. Movement update in this period succeeded in launching rich Japan with a strong army with a removable expansion. During the second half of the nineteenth century everything was in imported from the West Characterized Western technology and the Spirit of Japan. Japan was celebrated in 1996 that there was no illiteracy in Japan and the illiterate become now in Japan that is not fluent in the computer. Japan became a demilitarized state under U.S. custody and U.S. occupying forces tried during the five years of occupation (1946-1951) to change education system

.

In general, that this experiment into a new period of occupation was without military rule, the adoption of the peaceful route, not a clash of building development, has maintained its uniqueness through dialogue and maintain a set of traditions and heritage.

The Ministry of Education in Japan has set goals through the topics of National Education in accordance with an integrated system that participated all institutions of society and the interest of leaders at all stages of technology

.

The Japanese technological system used its capacity and strength in spirit and correction and to adopt such stringent qualities of the Japanese nation, which provides models for the people of national service and beyond any individual effort, as well as that Japan has not much affected by the West except for a limited period and then went out to establish a scientific base technical, industrial, professional direction on the preferred to theoretical studies.

What distinguishes man from other creatures being created future, since ancient times, the person is thinking about his future, but the tools were primitive but not serve his ambitions, but did not stop him from future search, then become the tools necessary to study the future available. The scientific progress in the sciences such as chemistry, life sciences, mathematics computer science and other sciences increased annually and became the basis for most other sciences. For example, enormous developments that have taken place in chemistry and life sciences were very clear and flourished in the new branches of chemistry, has seen tremendous developments in the study of atomic and molecular structure such as the use of lasers, X-rays with high energy, low-lying, gamma rays, the use of sound and light together in the study of atoms and molecules and ions in cases in gaseous, liquid, solid.

Very recent development in analytical chemistry included the diagnosis and recovery, as well as in biochemistry has emerged from molecular biology and life chemistry "Chemical Bionic", which has as far as the present and future, and received tremendous developments in the science of enzymes either in the life sciences, but taken with dozens of science sub-specialized anatomy and tissues, embryos and cell physiology, genetics, environment and animal and plant diseases and human behavior, sociology, fossils and geographical distribution of animal and plant classification and the date of the ancestors, the natural history of life and science of bacteria, viruses, parasites, worms, insects, fungi, fish, birds and mammals.

The Genetic Engineering and Biotechnology has used genetic engineering applications in medical, agricultural, industrial, production of vaccines, hormones and treatment of incurable diseases. An exciting scientific event was announced February 27, 1997. Namely it is the birth of the sheep Dolly by the method of transfer of a somatic cell into a specialized non-fertilized egg to another sheep after removing it nucleus and planted it in the womb of a third sheep. With other developments in pure science, the other is clear that the tremendous progress was hiding behind a backward in human cognition for other purposes, such as producing sufficient quantities of food to feed its human in the world as well as problems related to scarcity of resources and energy and the growing problem of pollution and population explosion.

The Arab brain drain Launched phenomenon specifically since the nineteenth century and especially from Syria, Lebanon and Algeria. At the beginning of the twentieth century, increased immigration, especially during the First and Second World Wars. Migration of doctors, engineers and Arabs scientists arrived to Western Europe and the United States, in 1976 to about 24000 doctors, engineer, 17000, and researcher 75000. In the last fifty years, the migration from the Arab Countries counted between approximately 25 to 50% of the total of Arab drain to the United States of America, Canada, and Britain - the two countries that attract more than 75% of the Arab immigrants. Moreover, 50% of doctors and 23% of engineers and 15% of researchers from of the total Arab competencies who graduated in the last fifty years now migrated to Europe, America, Canada and Britain, attracts more than 75% of Arab immigrants. Moreover, 50% of doctors and 23% of engineers and 15% of researchers from of the total Arab competencies graduating in the last fifty years now migrated to Europe, America, Canada and 54% of Arab students studying abroad returned to their countries. The Arab doctors working in Britain constitute about 34% of the total number of doctors working there.

The losses reached the Arab states as a result of the Arab brain drain to about 200 billion dollars annually, according to the report of the Arab Labor Organization in 2006. The Egyptian brain drain is representing big problem after the figures announced by the large body of statistics, which indicated that the number of Egyptians reached 824 thousand immigrants. From the competencies of Migration to Europe and the United States, Canada, Australia, the scientists represents more than 10 thousand immigrants and the rest in the areas of medicine, engineering, basic sciences, agricultural and humanitarian.

The report was cautioned by the Arab League in 2001 that Arab countries lost two hundred billion dollars as a result of scientific brain drain. The report pointed out that Western countries are the biggest beneficiary of more than 45 thousand Arab holders of certificates and qualifications.

The working paper submitted by the Department of Immigration and Population Policy in the Arab League for the first meeting of ministers was that the Arab migration has positive and negative results. The, countries of the Organization of Economic Cooperation for Development hosts million immigrants from Arab with advanced degrees, while five thousand doctors migrate annually to Europe.

The Arab Labor Organization issued a report in which it indicated that scientists are half of the immigrants to western countries, and the study indicates the reasons, such as the tyranny of political and government interference in the affairs of the universities of Arab States, and the lack of freedom of scientific research, underlying motivation for Migration.

Also according to the study of the Gulf Centre for Strategic Studies issued in 2004 the western states capitals attracted no fewer than 450 thousand of the Arab minds. The Arab countries contribute to 31% of brain drain from the developing countries to the west of capital by about 50% of doctors and 23% engineers and 5% of scientists. The memo distributed by Arabic Labor Organization, indicated that the past ten years have witnessed an increase in the number of immigrants of Arab scientists, especially in the fields of medicine and engineering and science, to work in Europe.

More than 2600 Egyptians working in scientific positions in countries such as America which came in the first place, while Germany came in the second and Canada in the third and Spain in fourth and France in the fifth. It is clear that Egypt is the biggest loser of the brain drain in various professions.

At a time when one study of the United Nations Development Program to that between (1998-2001), only emigrated more than 15 thousand Arab doctors. According to statistics issued by the United Nations Organization that almost 50 percent of doctors and 33 percent of engineers, 15 percent of the total drain of scientists from the Arab graduate, 70 percent of migrants attracting the United States, Canada and Britain.

The Iraqi scientific inefficiency leakage and migration at these rates reached maximum levels after the events (1991) and became higher education. In that era the development led to the departure of scientific goals followed after the migration of large scientific events (2003) and especially the year (2006).

The Arab countries, measures to curb immigrations and in particular Iraq and Egypt by legislation and measurement to reduce the brain drain, including the re-adoption of Law No. 189 of competencies of 1975. But the calamities that afflicted Iraq during eighties and nineties and after 2003 led to the leave thousands of Iraqis from the country's skilled and scientific qualifications.

However, the methods of dealing with the file of brain drain in general led to provoke the drain itself. Some countries have poured extensive privileges to these competencies, wishing to return to serve their country and possibly met with initial success of these invitations, but did not continue in its competence. To the opposite side, China clearly considered a lot of qualified scientists worldwide, in the framework of a political project.

Chapter Three

Patterns of Technology

Contents

. Preface
. Technological option
. Information technology in Iraq
. Communications and satellite technology
. Technologies of analysis in clinical laboratories

Preface

traditional definition of Technology includes the application of science. Technology is the sum of the ways and means based on scientific and experimental knowledge that are used in order to obtain a certain result, that is a productive knowledge that have solutions to discover tools or machines as opposed to theoretical knowledge and therefore requires the necessity of international cooperation in science and technology to serve humanity.

Since the early sixties, the United Nations addressed the interest in technology to carry out the development of growing countries and then held the in 1963 first conference of the application of technology for the benefit of the less developed regions in the framework of UNCTAD for the purpose of encouraging scientific research and the development of organizations and scientific societies in view of the role of technology in the rapid change quotient as well as scientific and technological revolutions. Since the end of World War II and the specifics characterized phase wares contradictory and restlessness and turmoil away from the stages of the previous history .

The reciprocity of different countries, becoming a dominant despite the considerable openness of scientific countries and due to the fact that science and technology is not limited to satiate the desires of scientists and researchers nor the human dimension and the social dimension . In the solution of problems and it requires a sustained effort to create a new international system depends on the basis of taking into account the current and future problems of the world fairer. Infrastructure spread over different sectors in the community and can be viewed according to two states: first, reflects the field of scientific activity and technological such as human resources development and scientific research and technological development , transfer technology, marketing and services (standardization, information, consulting firms), and the second: the level or depth or complexity State in which it operates in every area of activity ranging from level to level ended by goods.

There are of course external influences govern the work and performance of the technology system as it can cripple full work or haste and support . These influences are the presence of of supported policy ,strategies and mechanisms for the implementation of this policy by the state, and also of laws and regulations adopted to facilitate and push the system's work and the subsequent initiatives and national projects, and also of foreign and economic policy of the country, and the extent of their interaction with the global technology system.

The existence of technology system with a credit policy have positive effects linking each other and this makes the system into a national system of creativity or innovation. Technology is sometimes used to describe a particular technology and aims of the specialized techniques to specific targets and specific applications, , their tools and means to achieve these goals. The engineering profession is responsible for a lot of modern industrial technologies that enable people to live better Unfortunately, the small number of the world's population have technology and monopolize all the advantages.

The disadvantages of modern technology also accompanying some of the side effects of unwanted, has spread dramatically in the advanced industrial countries. Examples of these disadvantages are water and air pollution also contributed to technological development in the production of larger quantities of the most deadly and destructive weapons. There are different kinds of technologies, including communications , technology industry, space technology ,technology medicine , technology Engineering and Educational technology employs in various fields can be harnessed in other areas and there are examples such as the ☐communications technology / are employed in several areas such as medicine, media, economy and others.

☐ space technology / and are utilized in a number of areas such as education, media, manufacturing and other war.

Technological option

The technological option for development and change, and specifically technologies of knowledge that is the most important factor in the production of knowledge, which provides a logic technology, and then access to the production of more advanced technology. The developments in technology and communications take the pattern of open school education, due to adoption of the techniques of knowledge and information. Given the rapid development of societies and the transition to a knowledge society, will expanded greatly in the introduction of this type of education, with the justification that we have mentioned previously, including:

There is a clear relationship between the knowledge of the society and technologies, both of them are two sides of one coin. The role of the knowledge society is clear and shaped as a result dimensions of the dramatic advances in applications of technology, communication has evolved dramatically due to technology evolution. The technology turned to a revolution represented by one of the creations that has completed by the human being at the end of the twentieth century and early twenty- first century. Then this session expanded in the areas necessary for humans, including:

Contribution to a fundamental change in the pattern of human contemporary life the mind and thought became the basis of profit and investment. So there is sufficient justification which emphasized the importance and the need for Iraqi centers to take this type of education

To meet the challenges of a society of learning, attention should be emphasized on information technology and building knowledge base, networks of modern communications, the integration of technology in teaching, learning and research. The success of these education institutes and turn them into learning societies depend on the degree of the development of faculty cadres professionally and adopt the principle of participation and planning, activation and use of contemporary technology transfer the classroom into active effective learning environment.

The challenges of the knowledge society depend on the nature of the information and knowledge that are published daily in various parts of the world and at high speed, knowledge of different types. Including the globalization of knowledge, virtual of knowledge, and knowledge technologies. The globalization of knowledge based part of the community of the twenty- century atheist and named the third millennium.

Information technology in Iraq

The educational institutions could not build sophisticated information systems; provide internet services for staff and students alike, but that the ratio of students to the number of machines available is still very high. There general framework or the educational plans in information technology indicates the existence of chaos in the development of these plans that are in most cases.

We must set the general framework which defines the requirements and competencies to be developed. Educational institutions suffer from a severe shortage in the number of specialists from the holders of master's and doctorate of science and computer engineering as well as the software industry, which faces many difficulties and challenges that limit their economic feasibility and benefits of the future.

The entry of the computer in the field of education puts education planners in Iraq to the brink of social and economic changes as well as educational changes that must be absorbed and dealt with to create a modern educational systems. Then it can keep pace with innovation and development, and the number of computers in the Iraqi centres about (1000), where lost mostly during and after the war and the lack of networks of the internet widely, and in some centres and colleges have lost their advanced networks quickly and urgently.

knowledge society replaces the real values and standards of a new hypothesis occupies a distinct position in the knowledge society assumption; represent a model of high proficiency, difficult to isolate it from reality. The virtual universities are academic institutions aimed at providing levels of high-quality education for students in their residence places through the world wide web of the internet where these universities work to create an electronic educational structure, meaning that there is no need for classrooms or buildings or student rallies in classrooms or exams .

In other words, these universities established ranks of virtual default continue to conduct tests on after the adoption of advanced programs and therefore it requires Iraq to take the experience of virtual universities in view of what can be provided by funds, where low-cost student and regular pressure on universities. The virtual universities create platforms for a variety of electronic and administrative services and create an virtual science.

The negative contemporary universities problems are inherited, while positive changes create a lot of intellectual, cultural, social trends as well as the contemporary university job to prepare manpower and frontier science and scientific research, cultural and intellectual heritage.

Academic disciplines have evolved with the development of sciences and various new disciplines such as engineering, agriculture, science and total treatments that began in the nineteenth century, whereas the twentieth century indicates other disciplines such as business management, journalism, information and library science, economics, politics and world affairs were added.

Each country has special methods to determine its own disciplinary university and identification numbers, graduate students and the quality.

The world witnessed in the twentieth century breakthrough in all fields and scientific trends, so there are no boundaries between different disciplines. For example, medical science requires engineering science and recent tests of modern science depends on the physical, chemical extraction and analysis and also relies on mathematics to lay the groundwork mathematics.

The progress and development in pure science, for example, develop new subjects and disciplines and specialties of science. New interfaces were not known during the first half of the last century. The results of these major changes in curriculum and build up research transformed these developments to the university curricula. Seminars and researches are now carried out in different ways.

Scientific research base as the foundation for development is divided into developed and developing countries, including investment in scientific research. Therefore, it requires development of plans for internet in scientific research and development for the purpose of narrowing the gap with advanced countries by allocating appropriate annual budget of up to 1% of national income available through the effective administration and legislation.

Scientific research has contributed to changing economic and social features of the peoples of the world and the contemporary world to total countries vary according the level of economic, educational and scientific.

The university has a role in scientific research and has a high reputation in scientific research and even some of them have no good scientific research. In the United States of America there are around 3500 university and institutes, half of them carry research and teaching together and the other half mostly teaching scientific research is necessary to raise the level of university teaching.

Iraqi universities are able in very limited ways to build information networks that can be developed to provide Internet services to workers and students. The ratio of the number of students to the number of computer available is still very high and where there is no general frame work for the university plans in information technology is noted. A state of chaos exists in the process of developing these plans which are often duplicated. In the light of randomness must be a general framework that defines the requirements and competencies to be developed.

In (1964), 119 countries approved at the general conference of UNESCO, the recommendations that include the different forms of education outlying school and adult learning must be considered an interal part in the education system, the opportunity for males and females to continue to lifelong education. Patterns of continuing education in Iraq universities remain a valuable and suffer in Breach of the continuous weakness in performance, despite the long period exercised.

Virtual Informational Technology (IT) Knowledge

. Therefore, virtual community knowledge society replaces the real values and standards of a new hypothesis occupies a distinct position in the knowledge society assumption; represent a model of high proficiency, difficult to isolate it from reality.

The virtual universities are academic institutions aimed at providing levels of high-quality education for students in their residence places through the world wide web of the internet where these universities work to create an electronic educational structure, meaning that there is no need for classrooms or buildings or student rallies in classrooms or exams. In other words, these universities established ranks of virtual default continue to conduct tests on after the adoption of advanced programs and therefore it requires Iraq to take the experience of virtual universities in view of what can be provided by funds, where low-cost student and regular pressure on universities. The virtual universities create platforms for a variety of electronic and administrative services and create an oasis of virtual science.

Distance Education

The distance education system is an educational model based on modern technology, but there are many obstacles facing the problem of application, such as, recognition, criteria, measurement, standards and quality and others.

The positive results on the importance of distance learning as model of education compared to other models, suggest thinking in different forms and regulations and renewable methods of distance education to suit the point in time of the third millennium.

Adoption systems of distance education through computer networks on the concept of the overall approach include a series of general educational measures in electronic form so as to be accessible to scholars. Distance education is one of the important means of communications and technological revolution in the transfer of knowledge.

Its uses to develop and employ them in the development of human capabilities and enable environment for communication with the world of technology and information between individuals and among all sources of knowledge everywhere and produces the network of learner direct contact with science continuously. There are also regularly available information, photos, and recordings through the web, along with holding meetings and symposia.

The Ministry of Higher Education and Scientific Research with the American Academy of Science to convene a special joint program, to develop university libraries and to qualify Iraqi digital age, to bridge the information gap experienced by universities and securing sources of information more sophisticated to either academic researchers and graduate students. This information includes teaching and working in the rehabilitation of university libraries and to familiarize them with modern information technologies used in the libraries of major scientific institutions of academic science is   so-called virtual library.

Open University

The Open University appeared in Britain 19991 within the philosophy and objectives of the core functions of education, scientific research and openness to social surroundings. Some universities in the Arab Worlds have started since the fifties; the association is a kind of open study, because of the inadequate equipment rooms and laboratories. The idea of the Open University did not receive its application in Iraq and other Arab countries, except Palestine.

The Open University have resorted to innovative teaching methods, based mainly on self-education exercised by the students themselves not by studying books and educational materials that are registered by the Open University and through follow-up radio and television programmes broadcasted regularly in terms of appropriate information.

Communications and satellite technology

Communications satellites are able to grant broad prospects for the educational process at all stages of education, especially, in higher education. The potential has become possible to lecture from the university to house or transferred to another university.

University of Hawaii Islands began testing the use of satellite ATS- I to transfer voice message and printing between its various islands in the year (1971). Then established the university plans to send ground - receiving television broadcasts in educational exchanges between libraries and medical conferences, student and teacher training joint research.

In (1971) satellite ATS-I was used to provide medical treatment and educational programs for rural schools and some guidance to some medical lectures in the college of Medicine. University of Washington has also further tests at Stanford University in conjunction with Brazil using the moon ATS-6, as well as experience of the territory Rocky Ponte Moon using the ATS-6 and the experience of using the territory Appalachian Satellite.

In Ottawa, Canada, Stanford and California Universities in the United States, exchanges of experiences between teachers via satellite and mutual distances are often performed.

The experience of the Arab satellite ARBSAT from the Arab Satellite Communications Organization will be a successful solution to the problems of education in rural areas; also it will help to compensate and cope up with the serious shortfall in the preparation. There is no education of this kind in Iraq.

There is a clear correlation between knowledge society and technological knowledge, are two sides of one coin. The role of the knowledge society is clear and consists of dimensions as a result of spectacular progress in technology applications, where communications have evolved staggeringly.

Turning to technology revolution is one of the important creations performed by rights in the late twentieth century and the beginnings of the twenty-first century, then this session has widened in areas essential to humans,

To meet the challenges of community learning, it is required to work with attention to information technology, knowledge-building rules, modern communications networks, integration of technology in teaching and learning processes, and research. The success of these universities and transforming them into learning societies, depend on the extent of interest in the development professionally, and adoption of the principles of participation planning and stimulation of the use of modern technology and transforming classrooms into environments for active and effective learning.

The challenges of the knowledge society depends on the nature of the nature of the information and knowledge which will be published daily in different parts of the world and very quickly, knowledge of various types, including the globalization of knowledge and virtual knowledge and technology knowledge.

Iraqi educations are able in very limited ways to build information networks that can be developed to provide internet services to workers and students. The ratio of the number of students to the number of computer available is still very high and where there is no general frame work for the plans in information technology noted. A state of chaos exists in the process of developing these plans which are often duplicated. In the light of randomness, there must be a general framework that defines the requirements and competencies to be developed.

In (1964), 119 countries approved at the general conference of UNESCO, the recommendations that include the different forms of education outlying school and adult learning must be considered an internal part in the education system, and that the opportunity must be available for males and females to continue to lifelong education. Patterns of continuing education in Iraq education remain valuable and suffer in breach of the continuous weakness in performance, despite the long period exercised.

Students are able to learn through self effort, re-engaging directly by attributable steps to the year (1926), when Percy invented machine that include a series of questions and answers. Note: this education is missing in Iraq schools.

These elements are of good education to pursue promptly to search for useful work. They invest time, but this kind of study is not available n Iraq. Students from different educational levels are received, then provided with different types of studies required by developments of the individual and society. There is weakness of Iraqi universities to accept this kind of education.

Communications satellites are able to grant broad prospects for the educational process at all stages of education, especially, in higher education. It has become possible to lecture from the university to house or transmit a lecture to another university. University of Hawaii Islands began testing the use of satellite ATS- I to transfer voice message and printing between its various islands in the year (1971).

Then established the university plans to send ground- receiving television broadcasts in educational exchanges between libraries and medical conferences, and student-teacher training joint researches. In (1971) satellite ATS-I was used to provide medical treatment and educational programs for rural schools and some guidance to some medical lectures in the college of Medicine. University of Washington has also further tests at Stanford University in conjunction with Brazil using the moon ATS-6, as well as experience of the territory Rocky Ponte Moon using the ATS-6 and the experience of using the territory Appalachian Satellite.

In Ottawa, Canada, Stanford and California Universities in the United States, exchanges experiences between teachers via satellite and mutual distances are often performed. The experience of the Arab satellite ARBSAT from the Arab Satellite Communications Organization will be successful solutions to the problems of education in rural areas. Also, it will help to compensate and cope up with the serious shortfall in the preparation. There is no education this kind in Iraq.

Data Banks are information outlet to the problem of traditional storage and retrieval of information which can be stored in these centers. By using one of the elements described as document number, copyright, title research, objective data banks, we can link centers and scientific institutions and universities. These are tools of opening the doors in front of knowledge and information gathered by international experiences. In Iraq there are no techniques or terminology to this type of education.

Iraqi education have the option for the development and technological change, specifically technology knowledge, which is the most important factor in the production of knowledge and logic that provides technology and thus access to the production of more advanced technology.

The developments were strengthened in technology and communications to a pattern of Open Education, because of its reliance on techniques of knowledge and information, given the rapid development of societies and the transition to a knowledge society. Knowledge society has expanded greatly in introducing this type of university education.

There is a clear correlation between knowledge society and technological knowledge; both are two sides of one coin. The role of the knowledge society is clear and consists of dimensions as a result of spectacular progress in technology applications, where communications have evolved staggeringly. Turning to technology revolution is one of the important creations performed by rights in the late twentieth century and the beginnings of the twenty-first century.

Substantial changes in the pattern of human life become a modern mind and thought that is the basis of profit and invest. Therefore, there are sufficient justification that emphasizes the importance and necessity of Iraqi to take this kind of education, .

To meet the challenges of community learning, it is required to work with attention to information technology, knowledge-building rules, modern communications networks, integration of technology in teaching and learning processes, and research. The success of these universities and transforming them into learning societies, depend on the extent of interest in the development professionally, and adoption of the principles of participation planning and stimulation of the use of modern technology and transforming classrooms into environments for active and effective learning.

The challenges of the knowledge society depends on the nature of the nature of the information and knowledge which will be published daily in different parts of the world and very quickly, knowledge of various types, including the globalization of knowledge and virtual knowledge and technology knowledge will be dominating the world.

The importance of technological incubators

To support the private sector, these is a clear desire to encourage this sector and meet the urgent need of the Iraqi people, especially in the sector of pharmaceutical, medical, veterinary, and food industries, packaging, information technologies, means of education and industrial sector.

Elsewhere, there are efforts to link interaction of institution of higher education, scientific research and technical institutes with productions, services as well as benefit from the ideas of creativity and innovation among individuals and institutions of Iraq.

The embodiment of these trends has issued new laws and established special institutions that have national initiatives to achieve the required states of the initiative "cooperation mechanism" between the colleges and universities with various ministries and increase the number of graduate students, particularly at the stage of doctorate.

This shows that the economic climate and scientific activity and legislation in Iraq encourage the development of technological incubators at the present time (to be on a trial basis), and the objectives for that are the following:

- The first objective is to gain experience in how to achieve Iraqi "product development" or how to move from search result to the investment, or how to transform idea and innovations or developments and renovations to the scientific and technical institutions to factories or services. This experience and expertise that produces them, if successful, must be repeated in dozens of places and scientific and technical areas.
-
- The success of the idea of incubators leads to diversify the Iraqi economy greatly, especially when it beats the experience of being beyond generating tens of incubators in Iraq.
-
- The success of incubators leads to achieve substantial added value in production processes and services, not only in limited profits, leaving the added value of large foreign companies, the concessionaire or have intellectual property rights.
- Generating jobs and real productive for Iraqis, especially their graduates.
- Lifting returns of laboratories, equipment of universities, research centers and industrial development during the period of its life produced by investing in the work for production and service sectors and thus improve the conditions of their investment.

- The success of the experiment, solving these problems and constraints encountered will transform the experience of adoption by large companies in Iraq. Technology incubators associated with these companies will lead to significant results in the generation of existing industries (down stream industries) as well as in creating nutritious industries of these companies by generating (upstream industries), and feeding industries.

The technological incubators will benefit in the marketing of output of scientific and technical universities, research centers and industrial development and linked to the national economy more deeply.

Technology incubators in the Arabic world

The incubators are working to accelerate economic growth by supporting the establishment of the main engine of the economy and its dynamic and high annual growth rate. The establishment and technology that produce goods with high added value requires an incubator for the care of these companies. These incubators provide opportunities to accelerate the development of vocational skills for young people and specialists in the field of ICT incubator that was founded in Damascus between 2004 - 2006.

The Social Development Fund, adopted business incubators and technical mechanism to support the establishment of small projects and to develop the skills of self- employment among to initiators of technicians. Accordingly the Egyptian Society for small projects introduced incubators in 1995, approved the establishment of 30 incubators in Egypt, which then established the 9 incubators rely on simple technology in providing services, light manufacturing projects, which are dependent on knowledge and information such as the incubator of Mansoura, Asyout.

The technical incubators are located near or inside the universities and centers of scientific research. The technological incubator of Mansoura University, such as specialized informatics and biotechnology in the Mubarak city of Alexandria. Worth noting that one incubator accommodate about 40 projects to continue within Incubator for 3 years. Statistics indicate that 520 associate will enjoy the services of incubators in 2006.

To introduce technological incubator in Iraq it is proposed to bring these ideas systematically in line with the global approach by doing the following general functions:

- Absorbing the output and achievement within the country at the level of master and doctorate degrees and transnational consulting offices of these colleges.

- Absorbing the achievements of research and development centers in different quarters in the state (achievements of pharmaceutical, veterinary, programming and packaging materials)
- Employing the achievements of some companies from the public sector-private or mixed-dealing with specific technologies of the functions of these products, including incubators, and feeding of the proposed technologies as well as industrial products.

These technological incubators cover:

- Biotechnology
- Technology of drugs, medicine, veterinary medicines, herbal medicines, medicinal agricultural materials
- Technology of new materials, packaging, canning
- Information technology
- Technology of food industries

Justification

There are many justifications for the introduction of such regulation of such regulation, because the economic situation and its structure include more coordination, such as the following:

- Supporting the industrial sector in charge of the proposed technology incubator, including (medicines, information, food industries).
- Supporting the private sector which deals with these technologies to invest in the industry.
- Contributing to the provision of products for purchase of Iraqi citizens within its possibilities.
- Providing services not normally available.

Starting with incubators of this kind goes back to the physical possibilities available as well as the provision of appropriate venues as well as the fact that the process is not easy for the existence of many obstacles, yet this is a new experiment in which many elements have been evolved.

The private companies of special paper, cardboard, plastic, aluminum sheets, the National Center for Mobilization and packaging at the Ministry of Industry, conservation and packaging supportive of the pharmaceutical and food industries (dates, vegetables, and fruits) can play many functions.

Moreover, technological incubator in Iraq could play a special assignment, including:

- Linking universities and research centers in the industry.
- Marketing output of scientific and technical communities.
- Transfer of technology from home and abroad to invest in production and service sectors.
- Developing job opportunities for graduates.
- Serious desire to support the industrial sector and agriculture and the private sector.
- Generating companies in the feeder industries and expanding in the market and- added value and the generation of new industries.
- Increase the added value in production sector.

Specifications of technological incubator

The establishment of technological incubator in Iraq should be connected to the Ministry of Higher Education, to meet the urgent need for the existence of an institution that play a role of assistance and contribution in providing advisory services to medium and small- sized enterprises and individuals. Incubator package of facilities and support and consultative mechanisms can be provided during the period or periods of time until the relevant qualification is set to start production and actual work.

These incubators specialize in general successes of the features and services including:

- Medicines and Medical herbs.

- Medical and veterinary supplies.
- Informatics area, computer software and the Internet.
- Packaging, preservation, packaging (materials technology)

The private sector will participate in the sectors of pharmaceutical, food industries and information technology sector and other feeding industries and thus contribute and support the private sector in an orderly and controlled manner both in scientific and organizational transformation.

Number of international bodies and institutions, including the Economic and Social Commission for Western Asia (ESCWA), as well as UNDP, other international institution, regional, will also participate in this effort.

The mechanism of this incubator, include the followings:

- Development of training programs and consultations followed by the selection of leading scientific wish to begin work in establishing a yield of profits.
- Coordination of incubators and then choosing them from among enterprises.
- The incubator during the incubation provides financial services, legal advice and support and develops plans around the dilemma of funding and the necessary investments.

Objectives

The most import objectives of the incubator according to the proposed technologies (medicines, food industry, IT ... etc.)

- Helping graduates of universities and higher institutes to establish their institutions and their own business.
- Helping researchers to use the results of research carried out in technologies mentioned from the stage of laboratory work to the stage of practical application view production.

- Contributing to the resettlement of imported technologies and to assist in the transfer of technologies from developed countries.
- The incubator in the later stages to provide advisory services to the beneficiary institutions at work sites.
- The incubator provides advice in areas such as financial budgets discretion, and funding requirements needed to start production and organization of loans and payment methods.
- The incubator in the event of the availability of funds, provide soft loans for small enterprises.
- Incubators work mainly on developing a special relationship with local institutions, global development-related administrative and transfer technologies to local universities and research development.
- The incubator is usually implemented on intensive training courses for institutions incubated on some issues related to the success of the project.
- The incubator provides guidance on the new procedures and applicable laws entrepreneurs.

Those emanating from the other Ministries, including:

- Centre for the pharmaceutical industry.
- Center for Research and production of the diagnostic kits.
- Research Center for the veterinary medicines.
- Research Center pharmaceutical industry.
- National Center for Mobilization and packaging.

Therefore it requires the use of materials and instructions of existing laws to put the foundation or rules and procedure of the incubator from the laws we have mentioned above.

Moreover, passing laws may require the recall in the following basic features:

- Financial requirement of legal status and rights of individual property.
- Administrative requirements.

- Right of workers in the public sector when the loan was nurtured and how to leave his work in the sector.
- Issues of intellectual property rights for products.
- Legal features of the development of fund for incubator companies as they enter and when graduated and expanding production lines.
- Planning methods of benefit from the potential laboratory in the public sector and private sector.

The steps to implement the project

- Creation of a functioning business for the pioneers, examine the studying the legal aspects.
- Developing a plan of financing.
- Developing rule of procedure of the incubator.
- Agreement with the concerned parties.
- Organization of national symposium.
- Selection and training of nursing staff.
- Preparation of construction.
- Media and marketing
- Networking
- Follow-up performance

Funders

There is a number of proposed parties to finance this incubator project, including:

- Fund Development Planning Commission.
- United Nations Development program.
- Proceeds obtained from the 5% of the profits of companies that are allocated for research and scientific development.
- Cooperation mechanism.
- Incentives for creators.
- Consulting offices.
- Personal contracts for university professors.

Rules for admission to the incubator

The economic environment in Iraq determines some important rules to accept the products that will be adopted in this incubator:

- Product that has a relationship with proposed technologies that is currently marketing in the country (pharmaceutical and food industries ... etc.).
- Product that has the raw materials within the country.
- Product or service with high added value.
- Product that was developed inside the country (diagnostics kits, drug ... etc.).
- Product leading to the transfer of new technology (biotechnology, genetically modified organisms, genetic fingerprint).

Below is a list of preliminary candidates for certain technologies to adopt in incubators

- Biotechnologies, pharmaceutical, veterinary medicines, herbal medicines, medicinal materials, agricultural plants of dry areas.
- New materials technology, packaging and canning.
- Information technology, software, computer consumables, communications.
- Food Industries technology.
- Technology of electronic systems, control system, instrumentation, consumer devices, medical equipment.

Advanced Technologies

That the era of advanced technologies "High Technologies" or high-technology "Super Technologies" in which we live the last three decades of the twentieth century, the era in which we do not know how many decades it will take, representing a number of scientific areas and new technology comes on top of these technologies, laser and fiber- optic and space technology, new materials, pharmaceuticals, chemicals, minute nanotechnology, and finally biotechnology and genetic engineering.

The forthcoming technical applications that are difficult to know the extent of today and its impact on humanity can be viewed as the era of advanced technologies as the following day when mankind as a whole interconnected network giant relies on a wide range of communications satellites such as radio waves and X- ray laser, so that every part of the ground contact one of the satellites in the moment and will be available electricity in remote areas with farms, genetically engineered to convert sunlight into carbon and then to the crude stream, and can then run all the equipment and facilities, communications equipment, including satellite and the Internet.

The future applications of these technologies will be radical changes in the forms of life activities and practices relevant to the interests of individuals, groups and the process of coordination between these advanced technologies is a strategic way to bring about a surge in operations research and industrial beginnings began to appear, for example, a draft genome and bioinformatics. We will try in this article and subsequent articles offer examples of advanced technologies .

Femto

It means the number 15 and the chemistry of femto, to understand the reasons that lead to some chemical reactions without the other, one of the achievements made at the end of the twentieth century and the efforts of the world that have emerged Ahmed Zewail, who won the Nobel Prize in Chemistry in 1999 and showing the possibility of seeing how to move the atoms within molecules during chemical reactions using laser technology and the rapid use of a new standard of time is femto seconds .

Zewail has been used pulses of laser beam of a partial vacuum in the middle of materials to study the chemistry of high- speed stages of the transition, working within the Alfmto seconds be managed after the suddenness of molecules in the interim period and then became a pioneer of so- called Alfmto chemistry using laser technology (laser femto) camera and a very fast, sophisticated and very accurate to portray the ongoing chemical reaction between the molecules in three-dimensional image Alfmto time in its three dimensions, not one dimension only.

Finally, what scientist do is to identify cases of transition of chemical reactions as broken links and new links up, and the development of new chemistry carried the name Alfmto result of invention, or a new laser called laser Alfmto or laser technology and through rapid as we were filmed for the moment the chemical reaction within the atoms in the process of only one part of a billion a second, and therefore this technology and its owner, Dr. Zewail laser secrets complex world characterized by inventing something new the properties of new energy and knowledge of the movement of particles from birth or docking to know what was happening in record time is a million billionth of a second the proportion of this period to the second equivalent of one per second span of time to 22 million years ago.

To reach Dr. Ahmed Zewail of the use of laser microscopy to clarify the picture may have been the most difficult times in less than two and thanks to the time factor has been developed to see things, whether internal or external speed and one millionth of a billionth of a second.

The features of Applied Chemistry fmto side is represented as medical, industrial and agricultural in nature and changes in the human body, such as treatment of diseases such as cancer, diabetes, a cell can be imaged in the human body, and according to that disease can be determined in the light of the nature of these cells. That laser fmto which has been utilized for imaging the moment of the chemical reaction within the atoms in the cell process is not only is one of the thousand billion from the

Nanotechnology

nanotechnology, the technology that deals nanometer scale (metric unit of measurement), which is manipulating atoms for the manufacture of automatic equipment and information does not extend far a handful of atoms, and then anything can be made by micro- physics is quite different. The first part of the term to reflect the unit of measure (nano= 1909 meters)

the term nanotechnology in Engines of Creation written in 1986 and said the possibility of seeing the future of the armies of machinery hidden carry oxygen and nutrients and waste, and manufacture of atomic- sized machines called complexes pads that hold individual atoms. It was found that the control of the maize one and move freely and easily from attributes of nanotechnology.

This technique showed high density in the form of recent innovations in many of the global scientific publications. Including the mutations responsible for many genetic diseases and therefore provide in the future and relevance of information is indicated task to determine the organism is an information system for the manufacture of proteins and other compounds in spite of the inability to resolve the issue of structural triangular shape of proteins, despite the existence of mathematical models serve the purpose.

A fruitful field of nanotechnology research in many parts of the world and in government labs, commercial and academic, have emerged, according to the products on this technique such as sewing pants of fiber and manufacture of precision tennis balls retain flexibility.

In the near future, computers will appear smaller tubes made of carbon atom chips represent the atomic scale wires and high strength to build elevator to space, and plants that will manufacture computers minute integrated directly with the human brain to increase intelligence.

It is clear from the examples mentioned that nanotechnology is a technology that will change very little minutes every aspect of human life and giving people the ability to control the material, and this technique is the most important applications of medical treatment of human beings through the introduction of precision instruments within the cells to repair infected objects from within or for the diagnosis of patients as well as some developments on the mechanisms of control cells.

The first medical use of this technology has been developed a device implanted in the body that may manufacture. The energy comes biofuels in the cell and the purpose of this engine is the integration of machines in living systems fully, and use some of the scholars of this technology to produce nano- bombs to kill cancer cells.

A team of other customers of this technology to produce nano- bombs to kill cancer cells. A team of scientists of any other industry, the crew of siliceous teeth not larger than the size of the cell that can swallow the red blood cells and re- launched into the bloodstream, either antibiotics nano-particles "Nano biotics", which new types of antibiotics contribute to solving the problem of resistance of some types of bacteria to drugs, as well as the modified bacteria organisms, are converging nano- tube micro- rings 2.5 nm diameter amino acids and small hole walls of bacteria are infectious.

Researchers believe that the future of medicine is moving towards nanotechnology that will change medicine, as future devices that will work within the human body to diagnose many diseases and treatment. Russian scientists in the field of quantitative light and laser physics to reach a new discovery has been called the needles agency, a new type of X- ray beam, or special characteristics, as containing the elements of nano- any electronic material on subatomic particles that do not exceed the measurements of nanometer dimensions .

It has also developed the first computer chip companies auction that could contribute to increase the power of computers and a reduction, while reducing the amount of energy consumed by the chip is composed of cylindrical molecules of carbon atoms in diameter than a billion to a part of the linker carbon (smaller than a hair a hundred thousand times).

others technologies
Scans for Cancer

The specific type of nuclear scan you'll have depends on which organ the doctor wants to look into. Some of the nuclear medicine scans most commonly used for cancer (described in more detail further on) are:

Bone scans

PET (positron emission tomography) scans

Thyroid scans

MUGA (multigated acquisition) scans

Gallium scans

Nuclear scans make pictures based on the body's chemistry (like metabolism) rather than on physical shapes and forms (as is the case with other imaging tests). These scans use liquid substances called radionuclides (also called tracers or radiopharmaceuticals) that release low levels of radiation.

Body tissues affected by certain diseases, such as cancer, may absorb more or less of the tracer than normal tissues. Special cameras pick up the pattern of radioactivity to create pictures that show where the tracer travels and where it collects.

These scans can show some internal organ and tissue problems better than other imaging tests, but they don't provide very detailed images on their own. Because of this, they're often used along with other imaging tests to give a more complete picture of what's going on.

These nuclear medicine scans are commonly used for

Cancer

Bone scans: Bone scans look for cancers that may have spread (metastasized) from other places to the bones. They can often find bone changes much earlier than regular x-rays. The tracer collects in the bone over a few hours, then the scans are done.

Positron emission tomography (PET) scans: PET scans usually use a form of radioactive sugar. Body cells take in different amounts of the sugar, depending on how fast they are growing. Cancer cells, which grow quickly, are more likely to take up larger amounts of the sugar than normal cells. You'll be asked to not drink any sugary liquids for several hours before the test.

PET/CT scans: Doctors often use machines that combine a PET scan with a CT scan. PET/CT scanners give information on any areas of increased cell activity (from the PET), as well as show more detail in these areas (from the CT). This helps doctors pinpoint tumors. But they also expose the patient to more radiation.

Thyroid scans: Radioactive iodine (iodine-123 or iodine-131) is swallowed. It goes onto the blood stream and collects in the thyroid gland. This scan can be used to find thyroid cancers. Radioactive iodine can also be used to treat thyroid cancer. MUGA scans: This scan looks at heart function. It may be used to check heart function before, during, and after certain type of chemotherapy. The scanner shows how your heart moves your blood as it carries the tracer, which binds to red blood cells.

Gallium scans: Gallium-67 is the tracer used in this test to look for cancer in certain organs. It can also be used for a whole body scan. The scanner looks for places where the gallium has collected in the body. These areas could be infection, inflammation, or cancer.

Hemodialysis

The process of waste removal is called dialysis ,it is similar to osmosis .Semipermeable membranes in the kidney ,dialyzing membranes,allow small molecules such as water , urea and ions in solution to pass through and ultimately collect in the bladder ,then they can be eliminated from the body.

The most effective treatment of kidney failure is the use of amachine ,an artificial kidney , that mimics the function of the kidney .The artificial kidney removes waste from the blood using the process of hemodialysis (blood dialysis).The blood is pumped through a long semipermeable membrane ,the dialysis membrane .The water molecules , larger molecules including waste product in the blood and ions can pass across the membrane from the blood into a dialyzing fluids .

This fluid is isotonic with normal blood ,similar to in its concentration of all other essential blood components .The waste materials move across the dialysis membrane ,from a higher to alower concentration ,as in diffusion .Asuccessful dialysis procedures selectively removes waste from the body . Many dialysis patients require two or three treatments per week and each session may require one –half or more day of hospitalation , especially when the patients suffer from complicating conditions such as diabeyes .

Chapter Four

Technologies of analysis in clinical laboratories

Measurments techniques

The measuring of the concentrations of bio-molecules depends on the nature of their chemical structure, such as using iodine, for example, to measure the hormones of the thyroid gland and using the annular structure to measure steroid hormone.

Then many researchers use other ways to measure the concentrations of the hormones, including the use of antibodies, then both "Berson, Yalow" used the immunological method with radiation then the radio immuno assay". At the same time"Etkins" pointed out in 1960 the possibility of measuring Thyroxine by studying its association with protein " TBG" due usually to its presence in low concentrations in biological fluids.

The lack of precision and sensitivity of these ways has led to a lack of agreement of the amount of hormones concentrations (different methods of measuring hormones, life tests, chemical tests and immunological tests).

In light of the above, there came the need to find more accurate and sensitive ways then to was reached the design the basics of the Radio immuno assay.

Bioassays

The levels of PRL were estimated by using inconvenient bioassays, the classical of bioassays were based on the growth-promoting action of PRL using the pigeon crop sac or the rat mammary gland. There is fast proliferation of the epithelial lining of the crop sac, which contact with basophilic changes in the cytoplasm and they are sensitive to PRL, these changes were used as an indicated to the effect of the hormone.

Radioimmunoassay (RIA)

Radioimmunoassay (RIA) has been a popular method for clinical endocrine determination. RIA is a competitive protein binding (CPB) technique that uses radio-labeled hormone as the tagged hormone and antisera prepared against the specific hormone as a binding site. Competition between unlabeled hormone in the patient sample and the added-labeled hormone for a limited number of antibody–binding sites forms the basis of the assay. Although RIA is a good technique when high sensitivity is required

Immunoradiometric assay (IRMA)

These methods are similar to (RIA) in that a radio labeled substance is used in an antibody-antigen reaction. However, the radioactive label is attached to the antibody instead of the hormone (Antigen). In addition, an excess of antibody is present in the assay, rather than a limited quantity, is present in the assay. Because the entire unknown antigen becomes bound in IRMA rather than just a portion as in RIA, so that IRMA assays are more sensitive.

Both one-site and two-site IRMAs exist. In the one site assay, the excess antibody that is not bound to the patient sample is removed by addition of a precipitating binder, this binder is antigen bound to some solid support. In the two-site assay (Sandwich technique), a hormone with at least two antibody-binding sites is adsorbed into a solid phase to which one of the antibodies is firmly attached to the walls of the assay tube. After binding to this antibody is completed, a second antibody labeled with 125I is added to the assay. This antibody reacts with the second antibody-binding site to form the Sandwich comprising antibody-hormone-labeled antibody.

Hormones Receptors

The hormones receptors is divided into hormones membrane receptors "Membrane Receptors" and "Nuclear Receptors".

Hormones are working through binding membrane receptors through influential proteins accompany the membrane as active paths referring to "Second messenger signalting pathway" of "Cyclic Adenosine Monophosphate (cAMP)" and calcium "Calcium diglycerol", or other transmitters, that stimulates a series of enzymes the "Kinases". hormones bind to the cytoplasm before reaching the nucleus, such as the of the thyroid hormone receptors, which binds to the hormone in the nucleus without the movement hormone step.

Regardless of the class of receptors, there is a certain general principle of how receptors interact with hormones, associated with high affinity and high specificity, in order to allow access of a physiological response (functional) suitable.

The Bio response to the hormone is due to the displayed action on the target cell through its unions with relative specificity of protein nature called receptors, and these receptors received a great deal of attention because of its ability to bind hormone and the different their effects in bio chemical reactions that followed by binding of hormone to its receptors, such as glycoproteins, lipoproteins, neurotransmitters, drugs and several antigens.

Methods of assay of hormone receptors

There are three methods of assay for hormone receptors.

- The conventional biological methods

The conventional biological methods are used and usually set to determine the hormone sites in organs and tissues and include histochemical technique and Immunocytochemical" technique and "Autoradiography". These methods can not be used to measure the receptors concentration or get extra information on the receptors affinity and specificity.

- **Radioreceptorassay (RRA)**

The radioreceptor assay " RRA" technique is used to study the binding of labeled hormones with specific receptors in vitro. These studies are usually carried out "In vitro" and the basic needs of these methods the presence of radio labeled hormone and a source of receptors such as tissue homogenate , intact cells and membranes, or any of the parts inside the cell.

The hormone labeled with isotope is bound to its receptor until the equilibrium and a state of balance, then the amount of the labeled hormone receptors complex is calculated , and then the complex (labeled hormone - the recptors) is separating from free hormones by a number of ways including the centrifugation , gel filtration, dialysis at the equilibrium state, differential absorption and is then leaching "Decantation" .

- **Antireceptor- antibody technique**

A modern techniques which is used to study the role of the receptors to reach a physiological response and to find out the nature of the units for molecule of the receptors

Seperation techniques

Chromatography
Electrophoresis
Electrofocusing
Sedimentation
Spectral techniques
Diagnostic Imaging
- .Labelingwith radioactivity
- . Autoradiography
- "Membrane Filtration and dialysis"
  - . Protein engineering
  - . Immobilibzed enzyme technology
  - . Human Genome Project

- Bio- engineering
- Immunoassay
- Biotechnology

.Chromatography

Choromtography include multiple ways that all based on the separation of compounds based on the difference in migration through the passage in the center of force, as well as the tendency to face hard "Stationary phase" for central trans generational and face the hard nature solid or gaseous, or liquid depends on the tendency of various materials to hard to face multiple methods such as adsorption "Adsorption" ion exchange "Ion exchange" may include all kinds of chromotohraphy these methods.

Ion- exchange chromatography

Based on the tendency of the ions or molecules to materials other than for mobile and non- soluble, which owns the distinct shipments, or molecules that carry one or more of the positive charge exchange with the positive charge associated with the Ionia face, the mobile Resins "Resins" with a negative charge is called this ion exchange process with the positive charge "Cation exchange" and reverse ion exchange is called a negative charge.

Examples of the reciprocals of the ion non- animated "Immobile Ion Exchange" that are used in chemical research.

Polystyrene where will attend the multi- way styrene polymerization contributory "Copolymerization" with composite "Divinyl benzene", which adds to the styrene chains cross- shaped multi- written and added then aggregates the active ions altering the chemical composition of the original units that can be prepared for example, styrene resin that contains strong acid groups such as SO_3H hold process of "Styrene- Divinyl Benzene sulfonation".

In the same way can be prepared that contains the totals as strong as NR_3^+, or weak acid groups such as the $COOH$ groups or grass- roots groups such as NH_3^+ and types of preparation depends on the ion concentration reciprocals composite "Divinyl benzene" the amount of strings cross referred to the number listed after the name of the resin, such as "8X" Dowex 50, which contains 8% of the "Divinyl benzene".

Gel-filtration

The type of chromatography by gel filtration on the difference in the movement of compounds during the gels with regular pores partially used for the purpose of the way in a column filled with from one type of granulated gel filtration.

The pores is able to expel particles with, partial weights more than 10000 if we, for example at the top of the column a small scale solution of dissolved protein and molecular weight of 70,000 with ammonium sulfate the following happens:

Protein molecules expelled from the pores of granulated gel filtration.

The migration of protein size start "void volum" outside the granules column.

Interference ions NH_4^+ and SO_4^- small pores of the gel granules nomination so there is the amount of liquid required for the expulsion of these molecules outside the pores.

The granules are gel filtration where proteins are separated by major united ammonium by successive periods of time and be dependent on the size of the separation- free liquid, gel filtration is very important ways often used to separate proteins from salts "Desalting".

The gel gromoatography is the primary means of separation and purification of various enzymes and proteins, as well as the fragmentation of nucleic acids and proteins in the treatment, especially when quantitative diagnosis of some human diseases, as well as by the method of the exchange of tritium to test the protein structure or the structure of DNA and a study of the link between proteins and small molecules. The thin layer chromatography is mainly used for amino acids, polysaccharides and simple sugars, fat and various steroids and other small molecules. The ion exchange chromatography applications, including chromatography cellulose DNA, to purify proteins associated with DNA to separate in general the Bio-compounds according to molecular weights. The affinity chromatography is used to purify the enzymes and antibodies and transport proteins, membrane proteins and the chips and sugary proteins and the separation of animal cells in particular.

Electrophoresis

Most of the Bio- polymers with the electrical charge is transmitted in the electric field, which have the advantage to classify macroscopic particles and measuring the molecular weight and discrimination and diagnosis of amino acid changes with charges the components without charge, and vice versa.

The migration of the electric voltage low- lying useless to separate small molecules such as amino acids, either in the case of the migration of high voltage electricity will become more rapid separation of the amino acids, for example and using the electric gel migration and multi-acrylamide is the last for the time being one of the best prevailing circles to separate proteins and molecules with the addition of the compound. Dodecyl sulfate becomes possible to measure the molecular weights and nucleic acids and proteins, and at times used for this purpose the Agarose gel. The separation of DNA with a single when the use of acrylamide and Agarose.

The use of other electrical relay with a multi- acrylamide gel to separate the circular DNA which access through the gel. There are electric displacement by the immune system, which can separate the materials that have the same movement with vertical orientation or by gromotography followed by deportation cuts.

Starch gel used for the first time in the migration electricity and usually consists of' starch pastes, potato starch which burned grain thermally and after putting them in the organizer and the creation of the gel horizontally, as noted in the shape of the sample in a small part which consists of pieces of gel using a razor blade and sealed with wax is usually the part or material lubricating and then begin operation and pass the voltage specified.

Migration of the electric type of the SDS gel, it is possible to calculate the molecular weights of most proteins, the measurement of movement in transition in the multi- acrylamide gel, which contains "(SDS) Sodium Dodecyl Sulfate". In the neutral pH and concentration of 1% of "0.1, SDS" of mercaptothenal. Most proteins associated with multiple strings SDS and analyzed sulphide bilateral ties by "Mercaptoethanol". The result is damaged to the bilateral structure of proteins.

The complexes consisting of units of high protein and the SDS and imposes the existence and status of helical and random act of proteins of this method, as if with a regular form, have an equal proportion of the charge / mass, owing to the percentage of the amount of the SDS associated with each unit by weight.

Using samples of unknown molecular weight of two known molecular weights can then calculate the molecular weight of the sample unknown degree of accuracy between 5% - 10%, and this is certainly one of the most well- known methods and used these days to estimate the molecular weight of subordinate units. It is noted that the way the SDS gel can be used to measure the probability of presence of aggregates SDS. When an electrical relay to a series of proteins of known molecular weights by gel column first separated into a series of packets and then draw distance movement of the samples against the logarithm of molecular weight together there is a straight line.

Electrofocusing

proteins meaning they contain both negatively charged groups and positively charged, depending on the pH is positively charged when the pH and negative charge in the event that the pH is high. In addition every pH in which the zero- charge and the so- called point equivalent to the movement. For a mixture of proteins, it may point equivalent to multiple sites called points equivalent, which consists of a pH gradient in a column containing the tube- negative and positive. A wide range of points equal to the charge and the various compositions is a mixture of polymers of carboxylic acids.

Sedimentation

Sedimentation techniques is common in these days, which is used for the purpose of characterization of macroscopic particles. And using the appropriate type of these methods that can be measured by molecular weight, density macroscopic form of the molecule. It can also measure changes in these transactions and use one as a basis for separation of components of the mixture prepared for analytical purposes and preparations. Moreover, with the modern equipment measurements to make of high- speed centrifuge most of who has benefit. The main thing that occurs in high- speed centrifugal the movement of the minutes by which can measured the distribution of focus during the centrifuge tube several times and called the measurement during the movement of molecules through the center of the centrifuge, account for the molecular weight.

Characterization Techniques

Spectral techniques

In order to simplify the regulations of life systems, spectral techniques are used to study the structure of many of the synthetic compounds with important skill of life, including protein, nucleic acids and others. Moreover spectral techniques are used in the follow- up of chemical reactions of life and employment regulations of life in the human body.

The techniques continued often its development in medical diagnosis, which can be linked to any disease, accurate diagnosis of the variables affecting the livelihood systems and chemical structure of various compounds, or monitor the status of one or more of the fabric that differ in normal and pathological cases. Molecules absorb light and length of waves that are absorbed and the efficiency of this absorption that are dependent on both the structure of the molecule and its surroundings, making it a useful tool for absorption spectroscopy characterization of small macroscopic particles.

Spectrum of ultraviolet and visible, the measurement of absorbance in the ultraviolet and visible for several purposes, including measurement of the unknown substance test of some chemical reactions and diagnosis of materials and determine the structural parameters of the macroscopic molecules and follow- up the transition the coil for DNA, example is the double helix as well as the titration of pH spectrum of proteins and to identify features some of the proteins by way of solvent perturbation and by difference spectroscopy and identifying revealed the linking of small molecules to proteins and the union of protein - protein and solvent disorder of nucleic acids.

Applications of infrared in the Biochemistry applications of modern infrared for diagnosis of mutual hydrogen in proteins and diagnosis of the number of hydrogen bonds and aggregates and measuring the common effective group and broken through the process of metamorphosis and diagnosis of tautorserism forms by and the union between small molecules such as riboflavin and protein. In addition the identification the carboxyl groups in proteins as well as determining the status of hydrogen bonds in proteins and multi peptides to measure the direction of groups.

Raman spectrum
"Raman Spectroscopy of the important applications of the Raman spectrum in the compounds life, study the mechanisms of differs automerism structure and different kinds of amino acids and discriminate adenosine mono phosphate and triple phosphate and their ionic forms in solution as well as the diagnosis of helix, beta related structure random coil of amino acids and determine the number of disulfide bonds in proteins and determine the number of double bases in DNA.

Fluorescence Spectroscopy
There are two types of materials used in the fluorescence analysis of macroscopic molecules; the first contained the same molecules and

macroscopic materials fluorescent foreign insert. For example, there are three types of fluor of proteins self tryptophan, tyrosine, phenylalanine due to the fluoresce of the proteins resulted from these compounds, which could be used to study the changing perceptions of the enzyme by the positive correlation factor, assistant recipes enzyme active center and studies on the metamorphosis of the protein and the location of the tryptophars in the enzyme. In many cases the material could be added the fluor in the molecules composed as is being considered, either by chemical duplication or by simple correlation method.

The name of the method which will reacted by adding particles when analyzed by outer fluorescence. There are several requirements for materials and when the use of external fluorescence:

The material is strongly linked in a prime location.

The fluorescence must be very sensitive.

Should not affect the macroscopic features of the molecule that studied.

IR spectra

Measurement of the infrared spectrum between 4000 cm-1 in the upper left end and 925 cm-1 in the lower left end of the various groups (methyl, carbonyl, amide ...etc) and also to the totals effective adsorption characteristic frequencies to the group from certain areas and this range can be diagnosed by several groups of effective frequency and make the infrared fast methods and reliable.

The infrared spectroscopy in terms of a preliminary is not different from spectrometer for the visible and ultraviolet, where the device usually consists of three main sections:

The source of radiation the body which is heated to 1500 to 1800 Kelvin "K".

Radial Beam Analyzer, which is used to test wavelength.

Raman spectrum

Small portion located in the infrared in a non- extended with a frequency shifts, called (Raman scattering) and the reason for frequency is due to the high level of vibration by the addition of seismic energy to the molecule to electromagnetic wave optics.

In the process of scattering, the light is usually an irritating scattering (scattering) to the high level of vibration and lose energy and the frequency is decreased, on the other hand, when scattering center is located in the highest level of vibration (chattering in advance with the solvent molecules) can be transferred to the seismic capacity of incident light.

Raman spectroscopy uses the regular sources of the laser beam and light scattered through the Raman spectrometer using condenser of electro- optical and consists mostly of light scattered from the baseline formed either by absorption. The emission bred lines that compose a weaker Raman spectrum at lower and higher energy. And thus increasing the frequency and the use of normal temperature, where there is a vibrant and molecules other than the shaky, causing a decrease in frequency more normal, so the test fiber Raman vibrational transitions, such as infrared.

The Raman spectrum is not widespread and can be used to detect some effective groups. Some applications, including proof of structure for various kinds of amino acids, which is accompanied by distinct spectral changes in the support groups and carboxylic acids.

Fluorescence

It is possible that light energy absorb only when the molecule move from the lower energy to the top, such transitions in the diagram with vertical lines. When the molecule is initially irritating, it represent the excess energy that will be shaky as energy molecule in one of the levels of seismic and seismic energy appears as heat as a result of collision with solvent molecules (when the inflammatory molecule in solution and reduce the molecular, level to the lower vibration for S1).

In many cases, material could be added to the partial been studied either by duplication or by chemical simple correlation. Devices used to measure the fluorescence

A beam of light with high intensity it passess during the X- ray Beam Analyzer to choose the wavelength to the cause of irritation (eg, wavelength is absorbed by the material enough brilliant), after going through that light Irritated through cell containing the sample. To avoid detection the incident light is restored then the fact that emitted in all directions so that it can shine.

Dispersion of the optical rotation spectrum and spectra of circular birefringence "CD,   ORD"

A set of techniques that serve to know the status of the molecule or macroscopic and interactions, including absorption spectrum provides useful information of this kind, but that the study of absorbency of polarized light in a spectrum of optical rotation dispersion ORD and spectra of circular birefringence "CD" The method for measuring adoption of the wavelength on the viability of active rotation of polarized light and absorb the difference excellence polarized light and the direction right hand and the left.

The physical basis for each of the "ORD" and the "CD" are identical, and are in fact different to address the challenge of polarized light with optically active molecules. And cause to contain a very large part of the molecules of life on active duty visually Hence, the "ORD" and the "CD" has too many applications due to the fact that the spectrum of "ORD" and the "CD" of proteins and nucleic acids. The resulting from the spatial asymmetry of the components of the amino acids and nucleotides sequentially, but for the macroscopic particles it states each of the "ORD" and the "CD" in the structural studies of proteins and nucleic acids and proteins.

The pens and solids are used for this purpose at times, but the solutions are used in most of the time measurements for the "ORD" and the "CD".

Solution is placed in a container called a cell, and the device consists of the light source changes then the wavelength, and a system for polarization of the light, and system for measuring the polarization after the passage of light through the cell.

That the way to know the secondary structure of protein based on the measurement of the curves of CD, ORD experimental multi- peptide. As for the proteins they include measurements of the three main forms, alpha helix, beta form and the coil random.

It was found that the spectrum of proteins and the type that gives the same due to the impact of side chains on the rotary power by force, as well as peptide that occurs at times because of the hydrogen between two amino acids and peptides that have multiple heterogeneous long chain of the same strength turnover in each component, such as those owned by small cascade. The side chains of phenylalanine and tyrosine and histidine and tryptophan to the spectrum of CD when they are in certain situations and the sulphide bridges give two CD bands.

Recent developments in the application of circular birefringence CD and optical rotation dispersion ORD advantage of these modern methods to know the status of the molecule or macroscopic and the interactions between them, as well as the structure of poly peptide and it is believed that the key of the structural knowledge of the secondary structure of protein based on the ORD curves and the CD.

Furthermore, it is believed to use the test of changes on the status of the proteins by CD and study the changes in the structure of enzymes caused by the substrate, inhibiters that are related to enzymes, which can be illustrated by the CD spectrum for a variety of enzymes when the enzymes interact with the substrate and inhibitors of enzymes to help, as accompanying the metamorphosis of protein changes in the CD due to the loss of the structure of the alpha and beta and increase the spectrum of the components of the coil at random.

Nuclear Magnetic Resanace

NMR magnet is a spectral method that can provide sufficient information about the structure of bio-multicellular molecules and the interactions that occur between molecules, as well as molecular motion.

NMR spectrum depends on:

Displacement of chemical "Chemical Shift".

Fixed double "Coupling Constant".

Accordingly it used for the following applications:

Diagnosis of chemical structures.

Theoretical studies on the chemical tautomerism of displacement.

Studies on the tautomerism composition.

The dynamic characterization of chemicals.

Characterization of the spatial structures of chemicals.

Effect of solvents.

Technological developments of nuclear magnetic resonance technical developments have taken place in nuclear magnetic resonance study after the hydrogen nuclei (protons) and the nucleus of fluorine with high sensitivity and where the introduced technology transfer. Fourer the use of flashes of measuring the sensitivity of nuclei such as carbon -13 Bio-molecule such as proteins require nuclear magnetic resonance equipment with high frequency 300Hz and more, where it is difficult to analyze a device at low frequency (60Hz) or average (220Hz). New equipments of nuclear magnetic resonance frequency of 600Hz, these machines analyze many of the protein molecules and those of other techniques with the use of two-dimensional and three- dimensions.

New devices of NMR manufactured magnet after the entry of two-dimensional superconducting magnets, where the magnet save at -270 degree using liquid helium and liquid nitrogen to maintain the temperature of the helium and prevent it from evaporation.

To distinguish between one dimensions and two notes that the spectrum of NMR magnet with a one- dimensional recorded displacement and pairing constant on the same axis while the two dimensional displacement is recorded the on two different directions. The three- dimensional recording the two perpendicular to each other and are getting a nice Dox- spectrum. The most important uses of the latest is in the field of proteins.

Applications of NMR spectrum

NMH magnet is a spectral method that can provide sufficient information about the structure of molecules and macroscopic bio interactions that occur between molecules, which can calculate the arrangement of atoms in the spectrum, and the hydrogen atoms (difficult in a segregation analysis of X-ray diffraction can be identified on site by this spectrum), It is spread in the macroscopic particles, can also test different atoms (phosphorus, carbon, nitrogen and hydrogen) separately. These can be applied in determining the spectrum of protein structure and enzymatic study of active centers and link small molecules to proteins.

In general, there is a difficulty in analyzing the macroscopic particles, the presence of large numbers of lines of spectrum analysis, which distinguishes itself lead to difficult to diagnose due to the large numbers of potential tackles of each atom, where the border of the spectrum characterized by complexity theory. As well as would be possible to study chemical reactions and the effect of drugs on the disease and to study the changes that occur to the water molecules in living cells.

There is great potential for the study of chemical reactions using a spectrum of protein and phosphorus in living cells has made a technical developments in this study. The nuclei of other important life such as sodium, potassium, calcium, iron, cobalt and nitrogen. Examples of phosphorus- use spectral follow- up of the heart and life processes within the heart and to stop the interactions of life organic phosphates components as evidence of changes related to energy.

The specialists in the field of NMR spectrum nucleus of sodium, potassium and cesium due to the presence of sodium and potassium in the human body and play significant roles in the interactions that occur in the cell and pressure in the maintenance of fluid balance inside and outside the cells and the role of sodium in the transmission of nerve signals through the cells and the neurological contract and it can be completed as well as there are free and linked and contribute to solutions in the process of blood clotting and visual processes and muscle contraction, and here we began to record NMR spectrum of the nuclei of calcium and magnesium as well as nucleus of silicon, tin, nickel and other industrial significance.

It is played a great interest in oxygen -17 and carbon -13 as well as which gives information on the spatial structures and the use of nuclear magnetic resonance technique to the nucleus of carbon that is determined and these structures in detail. As for the oil industry played important roles in this technique and to identify the percentage of items to articles in different sections of asphalt, as well as technology that helps NMR magnet to determine the oil compound and petroleum products after the cracker.

Also used this issue to determine the quantities of oil in different kinds of grain linked to liquid plant cells without crushing and grinding grain.

The techniques of NMR magnet in the other areas are numerous and to identify the proportion of oils and fats, saturated fats in commercial fish, and set fat dairy products, cocoa and coffee industry and set the percentage of alcohol in alcoholic beverages. As well as to set the chemical structures of pesticides and chemicals in dyes and determine the toxicity and a lot of industrial materials. In the pharmaceutical industry it has been used a lot for confirmation of the purity of drugs that produced and the appointment of any impurities in pharmaceuticals.

A nuclear magnetic resonance other spectral method that can provide sufficient information about the structure of polymers and living on the interactions that occur between molecules as well as on molecular movement. The multiplicity of benefits and return to:

The possibility of calculating the order of the atoms of the NMR spectrum because it theoretically can provide information for the purpose of this account.

The hydrogen atoms (difficult in a resolution of analysis of X- rays), but it is possible to determine its location by magnetic resonance image.

Nuclear magnet line up in parallel magnetic field and the direction of high- energy or nuclear magnet lines up opposite field and can make a transition from two energy absorption quantity and energy of electromagnetic radiation and the appropriate energy to the heart of the nuclear magnet equipped with radio frequency range if we use the magnitude of around 10,000 chaos, but we see the resonance signals that nuclear magnetic protons that requires: Posted regular magnetic and the receiver of radio frequency. In practice, the necessary changes in the magnetic field bring the protons in the majority of chemical structures to the resonance is ten parts per million only.

To changes in the magnetic properties of some nuclei, particularly hydrogen nuclei (protons) and other nuclei such as $C13$, $P31$, $F19$, $N15$ and can be identified, as well as calculate the number of hydrogen in conditions of various electronic and more clear that can detect changes in the direction of nuclear in strong magnetic fields.

And can the chose in the different atoms (phosphorus, carbon, nitrogen, hydrogen) separately. And nuclear magnetic be very successful to determine the structure, when the use of small molecules. But when using macromolecules, it is difficult to identify large numbers of lines of the spectrum, as well as analytical distinguishing itself leading to the difficulty of diagnosing produced because of the line, but among the range of possibilities for every atom of tackles.

Using nuclear magnetic resonance, which will be important in the future to help and solve the problems of chemical and will be the difference between NMR and other methods when studying the macroscopic particles, where also be clarified the limited NMR characterized by complexity in the theory and devices that have been noted that many of these device had improved..

Nuclear magnetic requires resonance spectroscopy of the radio waves of the circular disk of the magnetic field and the absorption of radio waves recorded. The sample in the tube between the B- polarization of the magnet A load (104- 105 Chaos) describes the spiral shape in a plane perpendicular to the electric field of the magnet that surrounds the sample. The sender broadcasts reluctant high and constant (roughly 108 or circuits per second).

And the others in the magnet or in the frequency of radio waves in part cause no change in magnet strength, either the change in the frequency of radio waves.

Nuclear magnetic resonance capability to provide much information about protein structure, where the constants of the spectrum sensitive to changes in both the arrangement and status, for example, in the absence of any fusion or federation is a complex spectrum of peptide or protein which is the sum of the spectrum of protein components, which are amino acids.

Spectrum indicates the enzyme in natural forms and (random coil), which can be calculated from the amino acids of the protein, it is clear that there are significant differences between the spectrum which is calculated so that, it is noticeable, for example, noted the occurrence of poles that occurs when the large chemical proton in amino acid. But in the end, or carboxylic acids, or when the neighboring nitrogen atom to the bond, also a small composed displacement, when the proton adjacent to the carbon atom of the peptide bond, or carbon or nitrogen atom of the amino acid closest 2, also generated a small displacement, when carbon or nitrogen atom in the amino acid closest, in addition to this, when the protein status of normal displacement, spoke of the distinct chemical protons some acids in the form of security.

Diagnostic Techniques
Diagnostic Imaging

In medical diagnosis it is adopted mainly on the knowledge of diagnostic imaging technology spectrum, including the use of X-rays and gamma rays from, which is characterized by being electromagnetic radiation ionizing radiation, then began to think about using the term of this non- ionizing radiation infrared or microwave radiation and technical NMR magnet. The examples of spectral techniques used in diagnostic imaging:

X- rays
Gamma- ray
Ultrasound
Infrared
Anti electric tissue
Visual mechanisms

X-ray

The oldest techniques that is used in diagnosis and therefore will not focus on the importance of being where they were getting on the first pictorial representation of various tissues obtain after the development that is built on a limited computer assistance.

Gamma rays

The purpose of gamma- ray is the imaging profile then it was developed as computer- assisted also in the eighties which was called "ECT" and was then developed using imaging "Postiron emission tomoyraphy (PET)" where the radiation of tissue is carried out by position (positively charged) and thus can get a picture to clarify the life processes of the tissues that carry electrons and draw.

Ultrasonography

The speed of these waves are characterized by being less of electromagnetic waves, which provides an opportunity to measure the fetus as well as during the stages of development in the womb, added to that the fact that this technique is based on the fact that the X- ray is not ionized therefore it is not a preferred use in diagnostic imaging.

Nuclear magnetic resonance imaging

Despite this technology it is old, but it was then developed for the purpose of medical diagnostic imaging has gone from the seventies, where the nuclei of atoms is measured by the disposal of certain substances found in different body tissues. The criterion for the disposal of these seizures depends on the radio pulses that are similar to the frequency in the field of outer- core magnet and thus to obtain a diagnostic can be used.

In the medical applications for the purification of

nuclear magnetic resonance imaging to obtain imagery of infarction that occurs in some parts of the brain and important developments in this area the integration of multiple techniques and access to advanced apparatus for nuclear resonance imaging, including the "TMR" and "MRI".

It is important experiments that experiments are used the magnet resonance imaging of kidney transplantation, which was filmed nearby parts of the kidney and then infected the interactions that take place within the body after transplantation and efficiency of the cultivated parts. As well as imaging of tumors within the liver and liver imaging at the time of myocardial fibrosis or within, as possible, filming parts of the stomach and colon and to identify tumors. It was also to obtain information about stroke and is believed to imagery obtained of cancerous tumors of the brain were more pronounced than the use of X- ray.

It can be measured by any inflation occurs as a result of heart disease, and can also study the problems of the heart due to the presence of any obstruction or infarction in one of the blood vessels and could also portray the evolution of stroke, heart attack and its impact on the heart.

A nuclear magnetic resonance imaging "MRI"

This device is used which was created as a result of the development in the technology of magnetic resonance spectrum by the registration of spectra of life processes taking place within the animal body where the magnet- making with full slot by placing the human within the magnet and thus these devices provide a complete picture of the part which is conceived, and the advantage of the fact that this device magnetic field is not harmful, and the microwave radiation used is not harmful too.

It is possible through this device to study the effects of ongoing parts of the human body while taking a particular medicine can also be follow-up of the various core elements and sequentially, as well as to study the changes occurring stereoisomers of chemicals inside the cell as a result with other molecules. It has been known that nuclei , like electrons exist in different energy states or energy levels ,and the influence of electromagnetic radiation , transitions involving absorption of radiations can occur between the various nuclear states . This is analogus to the behavior of the electrons in an atom

.The nuclei of hydrogen atoms may be affaected in different ways depending on their positions in a molecule . These differences give rise to unique patterns of energy absorptions which is called absorption spectra , and the technique of nuclear magnetic resonance (NMR) has become useful tool for the study of molecules containing hydrogen. The NMR experimental technique was extended beyond small laboratory samples of pure compounds to the most complex sample possible – the human body .

The results of these experiments is termed magnetic resonance imaging (MRI) and this technique requires no use of radioactive substances , and is quick ,safe , and painless . A person is placed in in acavity surrounded by amagnetic field , and an image is generated , stored , and sorted in acomputor .Differences between normal and malignant tissue for example may clearly seen .

Electron Microscope

Electron microscope is using a torrent or stream of electrons, where the wavelength is too shoat then we can get on the ability of the analysis is very high. Extent of segregation (analysis) of the optical microscope and is an estimate of 2000 and this is not enough to see parts of the cell, viruses, and macroscopic particles, but the use of electron microscope segregation less than the uranium atom (in approximate) in special circumstances.

It is clear that electron microscopes large and complex and expensive operation that is similar to the foundations of the optical microscope, which reveal the sale electrons emanating from the source mail (metallic thread with a high degree preheated in vacuum) and reveals extensive by lenses and lens-body grows electromagnetic diffraction and finally drop the image as by the final lens of the projector.

To see the image on the screen is up brilliantly by the lens or can be scanned to imagine, as we mentioned earlier, the high segregation ability of this microscope to enable the researcher to view more details when you enlarge the optical microscope, the exact address. When the materials to be examined too thick for the passage of electrons then therefore requires the creation of a thin section, and it must be the sample used for the purpose of this solid and cut easily.

Nuclear techniques

Nuclear medicine imaging uses small amounts of radioactive materials that are typically injected into the bloodstream, inhaled or swallowed and is a branch of medical imaging to diagnose and treat a variety of diseases, including many types of cancers, heart disease, gastrointestinal, endocrine, neurological disorders and other abnormalities within the body. Radioactive emissions from the radiotracer are detected by a special camera or imaging device that produces pictures and provides molecular information.

Labeling with radioactivity

Require a lot of chemical analysis revealed small amounts of material with amount of concentrations $10^{-4} - 10^{-6}$ molari therefore it requires the development of other ways to respond to the concentration of low-lying, such as the development of experimental methods by radioactive to solve many of the other problems that might face them.

Some of these methods that could be used by dual- labeling for follow-up of two similar materials formed at various times by pulse method for follow- up fugitive substance at a time after the configuration without interference of other material. An example is the use of radioactive materials in the chemistry of life:

Choose a material that resides on small concentrations, which are difficult to measure by direct chemical methods.

Distinguishing similar molecules in different chemical sites.

Analysis of mixtures that are very complex, which can not be done by various conventional chemical methods. Including:

Enzyme interactions (DNA polymerase).

Measurement of molecular weight of the DNA by
labeling the final group.

Diagnosis particle by settling with the anti body.

Protein purification, which does not have a chemical test.

The isotopic properties will make the labeled compound more easily identifiable. For example, the radioiodine – labeled thyroxine molecules can be identified and quantified easily by virtue of their radioactivity .

The use of isotopes, both stable and radioactive, has proved great body of information in the medical scince. Stable isotopes are non radioactive and are suitable for use as tracers in humans. Especially infants children and pregnant women, stable isotopes have also been used in the quantative analysis of various substances in recent years.

Radioactivity measurements depends on the ability of radionuclides to produce ionized or excited atoms within the detector. Two basic types of radiation detectors are in common use: gas ionization ans scintillation .Radioisotopes allow the detection of minute quantities and differenate physically between substances.

The use of radioactive isotopes in biochemistry and clinical chemistry has proved us with a wealth of information about biological processes, that offers such as adiverse range of applications, using enzyme assays, biochemical pathways of synthesis and degradation, analysis of biomolecules, measurement of antibodies, binding and transport studies .

The use of radionuclide in nuclear medicine began when Frederick proescher published the paper entitled the use of radium for therapy of various diseases (1). Early experimental and diagnostic applications were performed with naturally occurring radionuclides, then the radioisotope with physical short half loves have become increasingly popular for imaging applications .

The first commercially available radioisotope generator was the $132Tc$-$132I$, (5) several other generators (such as Mo-$99mTc$, $68GE$-$68Ga$, $113Sn$-$113In$, $87Y$-$87mSr$....etc) subsequently evolved . These generators must meet certain physical basic criteria to be useful. It should. It should be simple and convenient to operate, radiation must be adequately shielded------- yield adaughter product of high purity in terms of both radioactivity stable contaminants during every clution throughout the life of the generator, the product should be in a chemical form suitable for use with amininmum of additional chemical or physical manipulation, lastly the radioactive yield of the daughter product during each elution should be high. Labeled compounds either be used in biochemical research --- routine medical diagnosis were carried out in vivo for medical diagnosis such as those labeled with gamma emitting isotopes to permit detection external to the patient, (12) but those labeled with beta-emitting isotopes such as: $14C$, $3H$, $35S$ and $32P$ were principally used in biochemical research .

There are various methods which were used for preparing labeled compounds such as of the followings:

1- isotope exchange reactions, in which one or more atoms in the molecule, exchange with atoms of the same element and of different mass, these atoms may be radioactive or stable isotopes, according to the following:

AX* + BX → BX* + AX

The compound BX under certain reaction conditions will exchange its X atom (s) with the compound AX* where X* atom (s) is an isotopic form of the element X. awide range of compounds labeled with different stable or radioactive isotopes are prepared by exchange methods, which have the advantage that they can normally be carred out on a small chemical scale. An example is the preparation of urea C, .

$CO(NH_2)_2 + {}^{14}CO_2 \rightarrow {}^{14}CO(NH_2)_2 + CO$

2- chemical synthesis in volves the construction of complex moleculry from simple isotopically labeled intermediates, yields are usually expre as a percentage radiochemical yield.

For example, the preparation of carboxyl-labelled fatty acids by reaction with the corresponding grignard reagent or acetic anhydride-14C, steroids 14C and amino acids-14C .

3- biochemical methods: these include different procedures such as enzymatic synthesis which isvery similar to chemical synthesis in that such aconversion usually occurs without any change in the specificity of the labeling or the molar specific activity (16). Total biosynthetic methods are normally of value only when microorganisms are employed, but the production of uniformly labeled carbohydrates by photosynthesis in detached leaves is an exception to this .

4- recoil labeling this method depend on the ability of recoil atom produced in a nuclear reaction to form a stable bond with an organic (or an inorganic) compound. For example, if an organic compound is mixed with a lithium carbonate or chloride and irradiated in anuclear reactor at fixed neutron flux, tritium compounds are produced by the recoiling "tritons" from the nuclear reaction $^6Li(n, \alpha)^3H$

One of the most radioisotopes used in clinical application in both cases as pure radioisotope or in labeled compounds is technetium-99m, which have a short half-life about six hours, with a predominate single photon gama emission having an energy of 140 kev. 99mTc-labelled compounds are diagnostic imaging agents used in the field of unclear medicine to visualize tissue anatomical structures and metabolic disorders. After interavenous administration 99mTc or it labeled compounds localized in specific target organ or tissue, can then be imaged using stable instrument.

TECHNETIUM CHEMISTRY

Technetium is not a naturally abundant element, some of its properties were produced by mendeleev in 1869, who called it ekamangabese and gave it the symbol (EM). After world war II, Perrier and segre gave element 43 the name technetium as the first artificial element.

To give rise to multiple oxidation states an forms coordination complexes with avariety of inorganic and organic ligands. The chemistry of Tc in its I-V oxidation states was sureveyd by davison and jones, many of the thermodynamically stable Tc-complexes have oxygen bound to it like TcO_4, $TcO_2.2H_2O$, Tc-O-Ligand (27-29). Tc V and IV complexes are knwn to have the ligand coordinated to the Tc alone, i.e Tc, Tc_2Ln....etc, the polyncuclear formation of Tc species is usually minimized by keeping the Tc-concentration as low as possible, to prevent the formation of Tc complexes with more than one oxidation state, efforts are made to control the reducing agent and the reaction conditions,.

These images can be superimposed with computed tomography (CT) or magnetic resonance imaging (MRI) to produce special views. Single photon emission computed tomography/computed tomography (SPECT/CT) and positron emission tomography/computed tomography (PET/CT) units that are able to perform both imaging exams at the same time.

Nuclear medicine also offers therapeutic procedures, such as radioactive iodine (I-131) therapy that use small amounts of radioactive material to treat cancer and other medical conditions affecting the thyroid gland, as well as treatments for other cancers and medical conditions.

Radioimmunotherapy is a personalized cancer treatment that combines radiation therapy with the targeting ability of immunotherapy, a treatment that mimics cellular activity in the body's immune system.

Nuclear medicine is used in the following organs:

Heart
visualize heart blood flow and function (such as a myocardial perfusion scan)
detect coronary artery disease and the extent of coronary stenosis
as bypass heart surgery and angioplasty
detect heart transplant rejection
evaluate heart function before and after chemotherapy (MUGA)

Lungs
scan lungs for respiratory and blood flow problems
assess differential lung function detect lung transplant rejection
Other Systems
identify inflammation or abnormal function of the gallbladder
assess post-operative complications of gallbladder

surgery
thyroid function , diagnose hyperthyroidism
evaluate for hyperparathyroidism
evaluate spinal fluid flow and potential spinal fluid leaks
Nuclear medicine is also used to evaluate the following diseases and systems:

Cancer
staging cancer by determining the presence or spread of cancer in various parts of the body
localize lymph nodes with breast cancer or skin and soft tissue tumors.
evaluate response to therapy

Renal

analyze kidney blood flow and function

detect urinary tract obstruction

evaluate for hypertension related to the kidney arteries

In children, nuclear medicine is also used to:

investigate abnormalities in the esophagus,...

...assess congenital heart disease for shunts and pulmonary blood flow

Nuclear medicine therapies include

Radioactive iodine (I-131) therapy used to treat some causes of hyperthyroidism and thyroid cancer

Radioactive antibodies used to treat certain forms of lymphoma.

Radioactive phosphorus (P-32) used to treat certain blood disorders

Radioactive materials used to treat painful tumor metastases to the bones

In RIT, a monoclonal antibody is paired with a radioactive material. When injected into the patient's bloodstream, the antibody travels to and binds to the cancer cells, allowing a high dose of radiation to be delivered directly to the tumor.

Autoradiography

This method is used to detect and locate radioactive materials in the cells or tissue for example, and so the molecule itself and is done by the impact of radiation emanating from radioactive materials or emulsions of photographic plates specially designed for radiation imaging device self-motivate, where silver halides grains, located in the emulsion as a result of the dissolution of radioactive materials in the sample, and the emission of radiation, including activation and work output reduction as indicators minutes for the site radiological effectiveness.

And signaling models resulting from the grains chemically and radiation efficiency in the presence of structures that are in contact with these granules and the microscope can be obtained from the resulting image on the two types of information at the site of radioactive materials and the quantity of a radiation of as the amount of silver particles is directly proportional to the severity of radiation present.

Of the modern applications of this technology as follows:

Measurement the number of molecules of DNA in bacteria phage.

Measurement the number of secondary units of the chromosomes.

Double vision in the DNA molecule of bacteria.

Other Techniques

Membrane filtration and screening "Membrane Filtration and dialysis"

(Like cheesecloth that has been used to separate the serum from the leaky). The cheesecloth filter extracts used in textiles. Then use the cards instead of porous fabrics for the purpose of controlling the size of chips and then create filters made up of cells, or glass yarn, either the softest materials they include sorting tubes membrane, which allows the passage of small molecules and ions. But keep the particles and macroscopic aggregates macroscopic particles. It is called the membrane tubes stitches the contrary, and that have the ability to separate the macroscopic particles from small.

Protein engineering

It is the technique that allows the installation of structural proteins desired in order to build a clone- mediated DNA "Cloned DNA". There is no relationship between the latter and engineering of proteins used, including the building of protein functionally, chemically and physically.

The DNA could be modified by two ways using:

Mutagenic in private venues.

Switch sections of the nucleotides.

The protein engineering include modify the structure with protein mediated by genetic engineering and most protein engineering is carried out currently in the field of enzymes, either to speed up its response to the incentive or to become more receptive to acid and heat.

Example: "Cloning" the cDNA for the receptor of "acetyl choline receptor" facilitated the technology which is called site directed mutagensis for getting sequences skilled "Deletions" or substituting some of the amino acids in an additional unit "subunits" of the receptor and then it can test these changes on the functional aspect, and are also defined as follows:

There are many examples of this type of modification for production of complex of organic compound that have catalytic activity have of it chemically synthesized for example the myoglobin of which associated with oxygen, but it docs not have catalytic activity. This Bio- molecule with three complexes of ruthenium "ruthenium" carrier of the electron through the surface of the histidines components generate a complex that has the ability to reduce oxygen and the oxidation of the natural ascorbate.

The construction of DNA contributed significantly to the development to the stage of protein engineering to construct proteins that do not exist in nature. The technique has evolved to the point can modifies the gene by an engineering to change the protein in a predictable and have to improve some functional characteristics such as:

No. transformation "turnover number".

Static Km of substrate specific.

Thermostability.

temperature optimum.

Stability and activity in non-aqueous solvents.

Privacy of interaction and substrate "Specificity".

Requirements of co- factors.

Protease resistance.

Allosteric regulation.

Molecular weight and composition of the structural unit "Sub- unit structure".

And for engineering the protein molecule, it is clearly necessary to ensure a series of rules relating to major synthetic building blocks of proteins that recipe as desired. After seeing the structural composition of protein crystals, it is then possible to diagnose those areas in which it occurs possible modifications to improve the catalytic molecule, protein, and this is done to modify the sequence of amino acids in the protein.

Major modifications protein

The use of site- directed mutagenesis determined then what is aimed to, because the change in one base in the gene result in a change in the sequence of amino acids in the protein, which in turn improve the protein in question. Large modifications in proteins by removing the "delete" section mediated by enzymes or by the unequivocal chemical structure of part of the gene. In this way, the production of spare "klenow fragment" "DNA polymerase" free of analytical activity, also can add sequence of amino acids through docking to improve the stability of proteins made in E. coli and finally can collect or part of a fusion gene or the whole of all or part of the other, thereby generating new proteins.

Determination the general features of the installation of the structural protein. Protein engineering based on the availability of information on the district and synthetic building blocks that are obtained from the methods of X- ray diffraction and nuclear magnetic resonance two- way "Two dimension nuclear magnetic resonance NMR" and the latter is the alternative method in the future. Many researchers expect success in engineering of proteins "Protein Engineering", especially after the great progress which has been in embryonic technique, where each protein is produced by genetic conditions of its own machine of the cell consisting of enzymes when they become three characters of the genetic material and arranged in advance and checked that then wrap as a specimen to be specific proteins effectively.

When you know the rules that allow the protein to form belts wrapped can then change the genetic information of proteins and identified so that it works in another way as soon as a large and powerful grants stability, and thus can benefit economically from the proteins of the broad areas of application by micro- organisms and can be more clear: for example, improved production of proteins (new physical properties and functional).

Important notes that are related to protein engineering is to clarify the potential relationship of proteins, where the protein for example, a specimen 15- amino acid. There are 103× 3 possible sequence of these acids is larger than the number of atoms that make up technical enzymes immibolized onboard, the development of these enzymes are restricted or limited to a solid surface to be in constant contact with the foundation to which the article in the mobile phase "mobile phase". It is clear from this that there is a possibility to use the many pathways that retains its effectiveness.

Protein that inhibit tumor growth solids tumors cannot grow larger than the size of pinhead unless they stimulate the formation of new blood vessels that provide the growing tumor with nutrients and oxygen and studies of angiogenesis,the formation of of new blood vessels , in normal tissues have provided new weapons in the arsenal of anti-cancer drugs. Angiogenesis occurs through a carefully controlled sequence of steps .In addition ,the growth factors and several proteins that inhibit the formation of new blood vessels and depends on the balance of stimulatory and inhibitory proteins .

Endostatin
isone of the anti-angiogenesis proteins and is a potent inhibitor of tumor growth and it binds to the heparin sulfate proteoglycan of the cell surface and interferes with growth factor signaling and as aresult the growth and division of endothelial cells is inhibited and new blood vessels are no formed .

Angiostatin
The role of angiostatin in the human body is to block the growth of diseased tissue by inhibiting the formation of blood vessels . There are about twenty angiogenesis inhibitors being tested in clinical trials involving humans .

Immobilibzed enzyme technology

At present, there are important industrial applications of immibolized enzyme technology represented by the following enzymes:

Glucose isomerase.

Aminoacylase.

Penicillin acylase.

Lactase.

The latter has been "Immobilized" on the particles of silica. It is used to convert the lactose in whey to glucose and galactose.

Applications to include of immibolized in the future as follows:

Use enzyme "Cholinesterase" for the purpose of pesticide detection "Pesticides" and watching the inhibition of this enzyme either by the method of electrical "Calorimetrically electrochemical" or by the color method.

Other enzymes that may be used in the same method in order to detect toxic chemicals, the enzyme "Carbonic anhydrase" is very sensitive to low concentrations of chlorinated hydrocarbons from low-lying "Chlorinated hydrocarbon" and "Hexokinase" to "Chlordane".

Immobilized diisopropyl phosphor fluoridate extracted from the nerve cells.

General aspects of enzymes immibolization

This process is intended as we mention it to determine kinetics of enzymes, as yell as cells that characterized by (desorption) on the surface such as fibers gels, etc., also can be used as phenomenon shooting accordingly.

Advantages of the immiblization process are the followings:

Finding the status of enzymes similar to those found within cells and tissues.

Prolonging the period of use and has repeatedly given to the survival of catalytic activity and stability.

Use appropriate concentrations and may be high for the purpose of increasing the speed of the reaction, given the focus to fit with the speed in specific circumstances of the reaction.

Contributing of the immibolization process to facilitate the purification process of related to products of reaction.

The use of multiple systems from the fermentation (continuous and open).

Reducing energy consumption and cost.

The immibolization methods are numerous, including:

Chemical methods: they are similar to affinity chromatography such as use the covalent and casual.

Physical methods: such as packaging inside a capsule adsorption and shooting.

As for choosing the appropriate method to be immibolized are determined according to the specific bases represented by measurement of activity, stability, so it must be taken into account the business side that is, have used with less expensive. And choose the easiest method because they are all tough and stay away from hazardous substances to human health, and the technical side is important in the selection process since there is a special mechanical pressure during the operation.

The immibolization cells vary from cell since it is being more of enzymatic system builders with the installation of diverse chemical content, therefore, requires that the appropriate modalities, simple and stay away from these that require to use extreme circumstances. It also requires that to taken into account the number of cells to be immibolized so the method must be a convenient and linking cells are good and avoid the use of hazardous materials. The characters of the immibolized cells are numerous advantages including the use of small amounts of carbon and energy sources and re- use of cells, so it is possible separate the growth phase from production phase, where it is possible control the fermentation. Immibolization depends on the type of cells, microbial cell reduce the size of the manufacturing process and thus reduce the cost of the production process. The Eukaryotic cells which are characterized as specialized capable of limited division of which are specific plant or animal cells and preferred to be immibolized, particularly those that are separated as any single and are generally used for the purpose of the immiboliaztion of adsorbed on the hollow fiber.

Enzyme Technology

The biotechnology is considered as one of the technical life in science and engineering. It was one of the enzymatic technology trends that have grown with the technology of life, despite being preceded by technical life, keeping in mind that enzymes from an engineering standpoint is a special case of the factors that have qualities such as privacy.

Bio- systems are used in critical periods in history to get the desired chemical conversions such as transformations of like milk to cheese and fermenting of liquids that contain sugar to alcohol, but such research trends have changed during the evolution of Biotechnology with the fact that these processes such as cheese, bread and alcohol industry still very important.

The history of enzymatic techniques started with the developments that have emerged a number of chemical transformations using the tissue of life, which include, for example hydrogen peroxide decomposition and degradation of starch to sugar and digestion of proteins.

Human Genome Project

The human chromosomes is 46 consisting of strips of the double helical DNA wrapped circumvent complex shapes of helix, normal and high and consists of the DNA with four units of high repeated synthetic (nucleotides incomplete oxygen), each of which consists of three components: nitrogenous base and sugar phosphate penta and not organic.

There are four types of nitrogenous bases and symbolized by TCGA arranged in pairs along the stretch of DNA and the numbers of secondary units in the DNA molecule. Approximately 3×10^7 base pairs in each of the cells of the human body, and the length of DNA equal 8×10^3 times the distance between the earth and the moon and bigger than the distance between the earth and the sun 300 times (the length of all the DNA, In the body 2×10^{10} km).

The gene (one gene) constitutes a piece of DNA it consists of a large number of secondary units and in the molecular weight of the gene is 600×10^3 and is a very long string of four characters, and each character represents a nitrogenous base. There are usually genes in the nucleus of the cell and molecular genetic consists of two bands are linked together by special bonds, each other on some twisting spiral and there is peace on the same wrapped that consists of a sequence of nitrogenous bases, or nucleotides that contain the bases arranged in a manner different from the gene to another and then discriminate organism from the other because of all the genes governing cell functions, guidance on ways making a specific protein or another compound with medical importance, hence a single gene responsible for the general one recipe and therefore we find that the qualities beauty, shapes and colors that each one of them result of a single gene or the number of genes.

Genes are transmitted from parents to offspring by mating the structural change in terms of affected and then the subsequent processes of making many of the compound causing the disease, which may be cancerous or always defect organisms.

The genetic map represents the order of genes (genes) within the cell chromosomes and that this arrangement within the human chromosomes is more complicated than other organisms. Thus, the process of discovering how to arrange these genes, given the sheer number and complexity associated with variation built and responsibility to control complex cellular functions. Hence the decoding process and diagnosis and scheduling of full human genome by genetic map and the preliminary draft of a preliminary genetic blueprint of human genes and the previous process is equivalent to a significant scientific breakthrough, scientific achievements made during the twentieth century, including the discovery of penicillin and landing on the lunar surface and use a computer and other discoveries.

And according to this perception announced 26.6.2000, the end of the main phase of the Human Genome Project, which represents the first achievement in the twenty- century atheist and the development of the draft map is almost complete and a preliminary blueprint for a human gene content of the human genome and was named the human genome. It had been prepared jointly by both the research centers m the United States, Britain, Japan, France, Germany, China and other countries with long experience in genetics and genetic engineering, funded by 18 countries.

.Bio- engineering

There are a number of scientific developments resulted from the diving in the world of molecules to push medicine forward through the discovery of technical of recombinant DNA (engineering life) and this new knowledge has led to the understanding of the causes of the disease that has eluded science until now, and thus to find new treatments to them. Engineering of life had an impact on medicine borders these have become easier with the forgotten youth of this important scientific field. The reality is that James Watson and Francis Crick did not reach a structural installation with a double helix molecule of DNA.

And then it was identified the gene (genes), which manages the production of individual proteins, and then we obtained the tools of partial strong, and in the early seventies researchers began snapped genes of the DNA. One of the species and planting it in DNA another kind for the manufacture of new molecules and in a few years researchers were able to transfer these genes and to produce objects that are within during the eighties and became a human gene transfer to many microscopic organisms and bacteria turning them into factories for medically useful proteins.

After it has been cloned of human genes in the micro- organisms for a number of hormones, including growth hormones and insulin in human as well as bacteria many of the genes responsible for human proteins with diagnostic value was produced at the level of marketing. It is noteworthy that human insulin is derived from living with diabetes, and also for the development of techniques for the production of antibodies "monoclonal antibodies".

Many applications, there is a steady increase in the use of enzymes in the diagnosis and treatment as well as in planting (farming) tissues and cells, "Tissue and cell transplantation" and that the development of engineering of life is still in the young stage, but there have major impacts on medicine and industry is synergy between electronic systems, electrical and life- component electrons so- called life "Bioelectronics" and electrochemistry of life "Bioelectrochemistry".

Then there have been the following design of a number of devices depending on what is stated in the above examples include "Glucose monitors" for the purposes of medical sensors and nerve gases for medical purposes and sensors nerve gases "Nerve gas sensors" to military uses. Based on sensors that have been most developed in the present time to reveal the exact products enzymatic activity mediated by the traditional pole "Conventional" where is the install (restricted) "Immobilization" new approaches that lead to devices with more sensitivity that depends on the movement of electrons between the direct- polarization and the redox centers protein "Protein redox centers" In brief the enzymes, which is based on the sensor depend on the medical sensor "Glucose sensor" and other sensors that measure chemicals in blood such as immune sensors include the electronic life "Bioelectronic immunosensors", which was commercially manufactured during the current decade, are measured in a large number of materials in the fluid of life, causing a revolution in the diagnosis, in addition to the incremental progress that has been happening as a result the development of a wide range of models "Sensors", which depends on the synergy between micro- organisms substantiated grants stability, "Immobilized and Stabilized". Finally, various data indicate that the microbiology of life through the engineering involved in the medical field in the production:

Antibiotics.

Vitamins.
Nucleotides.
Hormones.
Enzymes.
Vaccines.
Antibodies.

The progress that accompanied the engineering of life has affected in particular the daily practice of doctors, because of the speed that accompanied the evolution of knowledge and techniques in the laboratory and hence to the industrial production and then patient care. The expression of human insulin gene in bacteria E. coli, for example, has been studied in 1979 and that this insulin, with the original engineering- life of "Recombinant DNA" has been tested by volunteers with non- diabetes "non- diabetic" in 1950 and clinical trials that have been in patients with diabetes began in 1981.

The attention of most doctors on the applications of
modern life engineering in medicine, which tend to be very important in areas which have helped to revolutionize the diagnosis, treatment and understanding of many diseases, and examples of this therapeutically important protein, which was manufactured by engineered mediated microbiologist, microbiology, applications of single origin "Monoclonal antibodies", enzymes and others that arise out of uniform origin from lymphoid cells, where used in:

Treatment of cancer
Diagnosis of many diseases.

Containing anti- bacterial drugs that have contributed to engineering life and developed vaccines, hormones, vitamins and antibiotics and life for the purpose of producing these materials from micro- organisms after it was restricted to human and animal cells.

Hormones are the most advanced in terms of the accuracy of the technique used and the large economic returns through the engineering of life and led to great successes through the production of materials likes of the hormones which are stimulated, and stimulating the flesh wounds and the growth of the affected nerves that affect the sense of pain.

The success of engineering in the provision of life-hormones of the study and treatment has been a boom due to technical difficulties in extraction, which vaccine and growth hormones as well as the instigator of the secretion of pituitary adreno "ACTH" used to treat infections and diseases is used to treat wounds, burns, and stunting and release thyroid hormones pituitary as well as insulin used to treat diabetes, where possible transmission of their genes to bacteria.

Production of hormones is mediated by microbiology research center in the fields of engineering life in general and genetic engineering in particular, where microbiologists used to convert steroids and the production of hormones from the human body can not produce in sufficient quantities.

Then it was grown in importance after the custom of cortisone and its derivatives and their effective role in the treatment of arthritis, which draw many medical companies of steroids from plants, animals and chemical methods of trying to turn them into other steroid prescriptions. The methods of microbiology steroids is turning quickly but with less degree, there is in the addition of specialized microorganisms capable converting steroids quickly.

There is also the addition of specialized microorganisms is added hydroxyl group of any carbon atom present in the steroid. There are also some working to add hydrogen to steroids or withdrawal of hydrogen or oxidation or separate pools of chemical side effects. Using growth hormone that is released from the pituitary gland for the treatment of dwarfism find the hormone extracted from the animals be in a non-pure from, but according the production of this hormone is preferred to be extrated from microbiology such as the production from the bacteria E. coli after treatment genetically.

The plant hormones have been possible to produce from fungi, especially those produced from rice, as it is known that plant hormones industry is still expensive despite their limitations. In addition, there are a large number of proteins found in the blood such as the factors that contribute to coagulation missing by patients with haemorrhage as well as the albumin found in serum. These materials have been contributed to the development of production by engineering life in medicine (drugs).

The pharmaceutical industry, which includes anti- bacterial drugs, vitamins, vaccines and hormones of the biggest industries that relied on engineering techniques of life for the purpose of producing these materials from microbiology.

Medical applications of Bio- engineering

There are many faces, can be addressed when studying the medical applications of bio- engineering after the gene was designed, including:

Production of therapeutic: include hormones, such as somatostatin insulin, interferon and anti- biotic, where it was initially isolate the hormone somatostain for regulating secretion of growth hormone from the pituitary gland in the traditional way that requires half a million sheep brains to produce 5-10 mg of this material.

Treatment many of the genetic diseases: the treatment of many genetic diseases possible to treat many genetic diseases due to loss of protein production remedying these proteins from bacteria, and the examples of this case the planting and production of large amounts of genes to produce hemoglobin, which decreases in "Thalassemia" through the introduction of genes responsible for hemoglobin the patient's bone marrow, and then returned the cells to the patient.

Diagnosis of a number of diseases before birth: the fetus diagnosed in the prenatal stage, through the identifying the defects in a specific gene that causes the disease, such as some "Gamma- Globuinemia" and the disease lest Nhin as well as Tay- Sachs "Tay- Sachs".

There has been progress in some areas of medical engineering technology due to the recombinant DNA such as "cloning" the human insulin gene as well as growth hormone and its expression in bacteria that has been marketing of human insulin derived from microbiology and used for the treatment of patients with diabetes in addition to:

Production of interferon by a large clone human genes in microorganisms.

The development of production techniques and monoclonal antibodies and their uses.

The increase in the use of enzymes for the diagnosis and treatment in instilling the cells and tissues "tissue and cell transplantation".

Treatment of many diseases of genetic mediation by protein that being lost, which can be mediated by production of bacteria.

Diagnosis of diseases before birth by identifying the defect in a gene or several genes.

Turning to the relationship between engineering, medicine, is taken into account the following things:

Mutant cells and the cells unmodified organisms and their products such as antibiotic cellular life and plants, as well as other life transitions "Bioconversions".

Modified cells "Modified cells" and their products to ensure that objects Monoclonal "Monoclonal antibodies" of the following uses:

Immunological Studies.

Immunohistochemistry.

Tissue typing for trans- plantation.

Diagnosis and monitoring of malignancy.

Preparation of medicinal products with a "Prepartion of medically important products".

Recombinant DNA technology and its use for the production of insulin, interferon and growth hormone and vaccines "Vaccines" and enzymes.

The application of Bio-engineering techniques of molecular genetics and techniques diagnosis recombinant DNA in the diagnosis and (pathological) human disease:

Patriarchal diagnosis of genetic diseases.

Effects of genetic diseases on the specie disease.

Features of the future.

It is believed to that Bio-engineering represented by "Clinical biotechnology" has begun in the application management and industrial production of penicillin in 1940 that the success of the full insulin has created a growing demand for medicine (drugs).

The production of penicillin by fermentation and used in the treatment of diseases using the Bio-engineering problems that has been accompanied by the emergence of side effects and put some Bio-engineering solutions , and the problem of production has been developed through genetic improvement producing strains and control the components of the center other conditions contribute to the process of fermentation.

Bio-engineering and cancer

Bio-engineering has succeeded results in the field of cancer better than other diseases, as shown in the eighties that the main thing vs. cancer is a change in the genes (genes) from an engineering standpoint.

It was clear from the following entries in the relationship between the Bio-engineering and cancer.

Through analysis of a group of viruses called regressive "Retroviruses", which cause cancer in animals, as a number of these viruses carrying cancer-causing genes or tumor genes "Oncogenes". It appears that the retroviruses that cause cancer have been captured from the normal gene, cell, and one animal and made it part of their own genetic material. The retroviral infection of new cells in the later planted with genetic material, leading to the transformation of healthy cells into cancerous cells.

The researchers show that DNA extracted from human tumors can shift the cancer cells to cancer cells in test tubes. Or that a specific gene in a human cell that can transform sound cell into the tumor cell and a tumor- causing gene for bladder cancer in humans and called "ras" almost identical to the viral gene, a causing tumors in mice.

The gene tumor is often due to the mutant or increase in production and there is general consensus about the fact that any of the original tumor gene mutations may be some inherited mutations. The studies of funmor contribute to inherited breast or ovarian caner, the physician may be able to use that gene to assess the patient's condition and prospects and to provide more effective treatment for patients who have multiple copies of inherited suspicious. Harold Varmus and Michael Bishop has concluded that "Lancogen" the legacies of the genes responsible for causing cancer.

Bio- Engineering and AIDS

To understand the relationship between Bio- engineering and the AIDS requires a study of the topic in two cases:

How should the immune system to destroy virus: the defense forces resulting from the immune system to attack multi- directional and of different media for the virus (a specific target) to:
Phagocyte and other cells relevant to specific viral antibodies are chewing.

These cells installed in the grooves on proteins known as antigens of human white blood cells.
Construction immune complexes on the surface of cells identified by a type of white blood cells (T-help) "Helper T".

The recipients are on the T- cells help identify the peptide superficial "epitopr", associated with divide, and secrete small proteins that stimulate and activate T-cells and the toxic or lethal trait.

The killer T cells directly attack infected cells and fragmentation of viral particles and peptides associated with molecules of antigens of human white blood cells, when identified by toxic T cells by antigenic recipients on the surface of infected cells and destroy them by producing more of them.

The B- cells recognize the antigen norepinephrine viral surfaces as a prelude to their destruction.

Immune response and the virus "HIV" contribute the immune steps in defense against the virus "HIV", where they are:

Invasion of the virus of T- lymphocytes and cells assistance, followed by cloning and increase the virus and help decrease the number of cells, death, and loss of infected T cells.

Launch of viral particles from the cell membrane of T cells after being wounded by the T cells and B- toxic responses to be dispatched a strong defense which resulted in killing infected cells, viruses, and thus is determined by the breeding assistance and reference cells to a normal level.

A high level of virus gradually with the decline in the number of cells to help patients and reflects the so- called phase of AIDS when the number of cells less than 150 assistance cell in the blood followed by a rise in the level of virus with the decline of the immune system.

Monoclonal antibodies

The areas of application for the production of these antibodies where the potential for many therapeutic and diagnostic enormous, including:

Treatment of patients with leukemia and production of specific antibody alien objects on the cancerous blood cells, leading to the union of antibodies with and removed from the bloodstream.

Accepting the objects of a transplanted organ which are used Monoclonal antibodies or clone in the development of the body accept a transplanted organ such as the kidney.

Birth control through private industry specific antibody to proteins found in human sperm.

Determining the sex of the fetus through a special antibody to sperm of own unwanted sex.

Models are highly sensitive and privacy are being used as opposites, and a single origin and widely high sensitivity and privacy in early screening for malignant tumors by using specific proteins associated antigen and the presence of tumor presence.

Determining the levels of hormones in the body and used Monoclonal antibodies to determine the levels of hormones in the body and determine the effectiveness of the glands.

Search for the presence of some drugs in the body tissue and blood used Monoclonal antibodies in the search for the presence of some drugs in the body tissue and blood to prevent the occurrence of cases of poisoning or addiction.

Diagnosis of crimes using Monoclonal antibodies in the search also in the diagnosis of crimes. The food industry also used Monoclonal antibodies in the field of food industries, especially in the diagnosis and determination of the purity of food, processed meat, and free of unwanted substances and preventing fraud in this area.

Of the significant developments that have taken place for Immunology and molecular biology and biochemistry and the discovery of antibodies and the creation of a single origin "The Monoclonal antibodies" is characterized by privacy "Specificity" and sustainability of production, "Immortality" huge quantities "Large ruantities" and high purity "High Purity" for periods of a very long time.

However, these antibodies Monoclonal antibodies created by the multiple origin (clone) the molecular composition and effectiveness. Studies have shown that the use and applications of antibodies only be successful to detect very small quantities of tumor functions that can be used in early diagnosis of many tumors and by diagnosing the effectiveness of these antibodies could be argued that a large proportion of blood diseases can be categorized.

The advantage of imaging the immune flashlight as we have mentioned that the blue single antibodies prepared in the body of a patient associated antigen, surface of cancer cells without other cells and sputtering when labeling these antibodies with radioactive isotope, it can locate the radioactive iodine, for example by gamma cameras and thus can be located and the size of cancerous tumors, including colon, ovarian and skin cancer.

The unilateral clone in addressing some of the tumors where it can be linked to medicine as well as radioactive materials to these antibodies, such as chronic leukemia and thyroid cancer lymphoma and colon cancer has been found that these antibodies injected intravenously is grappling with the tumor cells and selectively and is disposed of, where became can direct these drugs directly to tumors by linking them to the catalytic antibodies to these tumors. Used monoclonal antibodies to treat cancer when there is a high toxic concentrations in the tumor.

It could also be linked Monoclonal antibodies radioactive isotope and alive in the body of a cancer patient at which time the radioactive material to the site of the tumor and therefore within the cancer cells and it crashed. There are many researches addressing the use of monoclonal antibodies in the early diagnosis of the body rejecting the case of the tissues and the transplanted organs as well as a lot of studies on the use of these antibodies in the treatment of the case of rejection.

Some applications objects Monoclonal

Improving the sensitivity of the current immune for tests or tests new Histocompatibility
Fibronnectin
Blood groups Antigens

Sperm antigen
Interleukins IL
Interferons
Progesterone gastrin
Blood clotting factors
Estrogen
Human growth hormone

Monoclonal antibodies has clear impact and important role in clinical medicine before developing the "hybridoma technology", which provides heterogeneous objects "Homogenous antibodies". The research carried out by each of the "Kohler & Milstein" in the early seventies, created a method used for the manufacture of the anti body homogenized with a quantity of non- specific proliferation applied at large.

The researchers "Kohler" and Mlesstin have participated in the production of monoclonal antibodies, which is derived from specific tissue culture which is called hybridoma, where the latter's has the ability to produce one type of antibodies but does not produce more. This is done by crossbreeding or mating types of cells, the first is produce the antibody and the second for the growth of cancer cells have the ability to reproduce.

And then treated with hybrid that has to be the formation of antibodies, where antibodies are produced for this body alone, and perhaps it carries the qualities of cancer, the production of antibodies is very large quantities. It is possible in the light of the use of a composition for the manufacture of an unknown antigen monoclonal each part, and then used these antibodies to probe the chemical composition of the real knowledge of the unknown substance.

Monoclonal antibody can be used for treating patients with cancer of the blood through the manufacture of these antibodies is specific to the alien objects on the cancerous blood cells united for the purpose of removal from the blood stream, and used these antibodies for early detection of the presence of tumor cells through the tests that require purity too high to measure the presence of proteins associated with its existence of these tumors and their locations in particular antigen-mediated tumor.

These antibodies are used in determining the levels of hormones in the body and to determine the endocrine events are also used in the search for the presence of certain drugs in the blood and tissues because of the poisoning have also been introduced in the diagnosis of bacteria in the development of the transfer of the body of a transplanted organ, in particular kidney.

Immunoassay

The development in many areas of clinical medicine by "Yallow & Berson", which developed a radio- immunoassay technique "RIA" for the purpose of measuring the concentrations of very low- lying materials and the offer of "displacement" antigen, which is marked by radiation from the body own by adding increasing concentrations of antigen, the record is marked by radiation and this applies well in the science of hormones, as the hormone levels in the bloodstream always be very low, making it difficult to measure ways of life and conventional chemical.

There are many hormones that can be measured easily and quickly test by radiation immuno assay including, prolactin, which was found to be associated with spinal tumor glandular "anterior pituitary gland tumour" and more permanent with symptoms by menstrual "Menstrual disturbance". In fact, the measurement become a part essential to the tests of modern futility "infertility".

Measurement by radio- immunoassay and other methods of link

There are three methods of test for the purpose of measuring the materials of life:

Biological assays.

Binding assays.

Physical chemical assays.

There are also two types of tests in cases of binding namely:
Test the link "Ligand".
Tests of Binder.

The test of linking section are all kinds of bands that can be used, namely:
Cell receptors.
Circulating binding protein.
Antibody.

The types of acquisitions "Tracers" used are included:
Particle.
Fluorescent.
Enzyme.
Isotope.

Applications of the main principles of radio- immunoassay test

Radio-immunoassay test depends on the competition between the antigen, which is labeled and non- labeled sites on the anti body component complexes ratio on the amount of antigen without radioactivity.

Immunohistochemistry

Using this antibodies tagged for the diagnosis and distribution of antigens by the optical microscope or electron microscope in tissue sections and sandwitch technique used widely. The unlabeled antibody is placed on the section washed with labeled antibody increases and the layer enzyme linked or marked by "fluorescence" against immune "Immunoglobuin" and can therefore be signaling antigen under examination.

This technique in applied research (Bio and medical research) Examples using antibodies "antisera ordinary polyclonal" on the "Topographical mapping" of the various types of cells in tissues such as "islets of langerhans".

The immuno tissue chemistry containing various anti blood serum describes beta cells of containing insulin, located in the central mass of the island "islet" cells while the A cell that secrete glucacon in the peripheral side linking them to the cells D secreting "somatostatin". An examples on these the use of monoclonal antibody in the clinical diagnosis of tissue "histopathological" of the disease when the test of the "Lymph nodes biopsy" where it help in the classification of a certain type of lymphoma "lymphoid tumour" (e.g Hodgkin's disease and various types of lymphoma "tymphoma").

Histocompatibility antigens

These antigens are characterized by being glycoprotein's present on the cells, especially white blood cells in humans and is then called human white blood cells, "Human leukocyte antigen" (HLAs). Carrying this antigen genes of the immune response status, where they control antigens were present in different tissues in the body, in addition to genes, there are mismatch humoral immune response "Humoral" and cellular "Cellular".

Immunodeficiency

Immunodeficiency resulting from the lack of a genetic condition in the inability to create an immune cell, or one of its outputs and symptoms of the disease- causing immune deficiency commensurate with the degree of destitution and accompany him.

One of the examples on the case the disease resulting from this deficiency (AIDS), "Acquired immunodeficiency syndrome" (AIDS) or acquired immune deficiency syndrome and was attributable to a virus of the type of regression "reterovirus" with a tendency to lymphatic cells "T" in humans, called "Human T cell lymphocyte".

This virus has several methods to spread such as blood and mucus and the interface is accompanied by injury to the virus to many diseases, so called on the situation of the disease and syndrome of one disease. The assumption is based on the perception of geometric depends on logic that a collapse of the immune system caused by AIDS, produced by relationship engineering between HIV (human immunodeficiency) in the body of the infected and the immune system where it is can do the following:

Cloning of the virus rapidly that destroy large number of cells of the system as there are two types of tests in cases of a binding.

Faces a viral reproduction for many years through every defensive response to prevent the virus from reproducing.

An imbalance in the latter for the benefit of the virus "HIV" event leading to AIDS.

It can show new geometric forms of the virus as a result of mutations that be able avoid the defense forces of the body in some way, and confuse the immune system, which enable many patients to stay healthy for many years, finally collapse due to the boom continued, speaking of the virus.

Preparation of medically important products

The use of monoclonal antibodies the purpose of purification and preparation of medical materials which is represented by the task done by the "Secher, Burks" and that covalently bound to anti- body unilateral origin against assigned to drive in "Sepharose" and therefore can refine 5000 times.

There are a number of the production of insulin- led company "Eli Lilly & Co", which used the Bio- engineering, including recombinant DNA technique as a base for the manufacture of human insulin. The production process has been carried out by "Lilly" in cooperation with "Genentech Inc." According to the following steps:

Determination the sequence of DNA From the known sequence of amino acids m insulin.

Chemical structure of genes for the series "A" and Series "B" of insulin, contains each and every one of them in the codon methionine at the end "O".

Each gene enter in 2 mentioned in the beta- 2 gene "B- galactosidase" of the plasmids, which are the same within the" E. coil".

Because of the fact that the bacteria had grown in the medium that contains the galactose and not glucose that urges enzyme B- galatosidase and then with a series of insulin "A" or "B" linked with methionine.

After the breakdown of bacteria, treated with cyanogens bromide "CNBr", that breakdown the proteins at the site where the methionin is present.

Purification of the two strings "A& B" and then returned to their union with the natural production of insulin, the two strings.

The bacteria do not carry enzymes that change the pro insulin to insulin through manufacture the lofty strings "A, B" in the bacteria followed by purification separate chains bilateral, ties with the sulfide.

It is noteworthy that human insulin that is produced by E. coli was tested by healthy human volunteers and with diabetes (and there were no adverse that there is capacity similar to purified pork or with decrease the blood glucose when injected under the skin or injected inside the vein.

The trials have been carried out on human and compared with animal insulin producer from the pancreas of pork produced then was shown similar effects. This was in the hospital, "Guys" in London and Osaka in Japan.

The department of food and drugs in the United States the U.S. approved marketing of insulin produced by the micro- organism, which called the "Humalin", also got the same thing by Britain in the same year 1982.

Growth hormones

These hormones are used in the treatment of growth disorders in children, dwarfism some cases of infertility in women and because of the high cost of treatment using these hormones and the difficulty of obtaining and the need for a large number of the pituitary gland the high cost of the hormone and the likelihood of infection with viruses and expected such as uses future growth of the tissues, and the flesh wounds after surgical operations the flesh of fractures and the assist once in the treatment o burns, sores and the used in the study of malignant diseases.

Researchers have made efforts to extract it and its production in bacteria by bioengineering after the successful transfer of genes responsible for hormone production from human cells to bacteria which was done in 1979.

The length of human growth hormone, 191 amino acid with a molecular weight of 2200 excreted by the pituitary gland, secreted of from the front gland of the longitudinal growth of the structure which means the isolation from the pituitary gland.

Interferon

The interferon is extracted from the cells infected by virus and studies have shown that these cells, the immune system stimulates the production of this hormone during the infection by virus.

Then overlap with the later injuries to the work therefore it is called "Interferon" used to treat the casualties.

The cost of purification of interferon is very high and extracted from white blood cells, and the other efforts made to develop the production method for human interferon non- blood through tissue culture (artificially), and then methods developed to produce interferon from bacteria, it is done successfully in 1980 and it became clear that there several species produce a number of genes.

Interferon has been isolated in 1957 and considered at that time the first line of defense against attack by viruses, used to treat many viruses diseases and including:

Cold.

Hepatitis.

Cancer Diseases.

Because of the interferon ability to prevent abnormal complications of the cells studies have shown that the immune system of these cells motivated during virus infection leading to secrete very active material is to overlap with the later injuries to the work.

The interferon's a family of proteins, discovered as a result of infection which flows into the cells by viruses "Virally infected ceils" are characterized by the following characteristics:

Antivural in other cells.

Inhibition of "Cellular proliferation" "anti cancer drug".

Internalization of the immune system "Modulation of the immune system".

It is possible to re- classify the life can be interferon's to the following types:

α- alpha- interferon leucocyte "α leucocyte interferon".

β- beta, interferon cell Fiber "fibroblast interferon".

Gama of lymphocytes, immune interferon "immune interferon", "Lymphocytes T".

Leucocyte interferon reduces the spread of a vesicular composition "vesical formation" and responds to fear of infection of the liver "Hepatits B" as well as in various malignancies "malignancies" such as breast cancer spread "metastatic breast cancer" and non- Hodgkin's lymphoma "Non- Hodgkin's lymphoma" and osteoma flesh "Osteosarcoma" and malignant melanoma "Malignant melanoma".

Research has shown that there are about 20 types of interferon, which produces a number of genes mediated for the purpose of genes engineering more effective against viruses or against tumors.

Biotechnology

The concepts of traditional genetics known for thousands of years and then evolved into the development of Mendel's laws, famous, and then changed, according to the progress of various technologies and discoveries.

The increased in gene progress in terms of chemicals, then was reached as to how the work of the gene at the molecular level with the adoption of methods of biochemistry, rather than traditional methods in the interpretation of genetics, which paved the way to the evolution of the concept of genetic engineering.

The high and low living organism from units can only be seen with a microscope, a cell, which contains the kernel and the last, which includes chromatin materials that turn into chromosomes (chromosomes). Studies show that both the egg and sperm contain half the number of chromosomes (chromosomes) and therefore half the number of genes in human egg contains, 23 chromosomes. Chromosomes chemically composed of proteins and the "DNA". Studies have indicated that the gene is made up of sections from the chemical DNA. The latter consists of chemically according to the double helix model of Watson and Kirk.

That has been proven that DNA is the genetic material according to the following:

The amount of DNA constant in all cells of the individual, regardless of the quality of tissue that make up the member.

The DNA ability to configure a mirror image of himself during the division.

The DNA characterized to contain all the genetic information in the order of succession of base nitrogen.

Genetic engineering

Genetic engineering is intertwined with the vitality and technology based on several basic sciences such as cell science, genetics, biochemistry, physics, and others.

The content of genetic engineering the human ability to control the mechanisms of gene transfer from one cell to another and how to express them within the cell for the future.

To understand the genetic engineering practice to be done the following:

Isolation of DNA of the object which is meant the transfer of its genetic material.

Cut the DNA to the sections that each end section to a particular gene.

Identify the gene required between these parties.

Ensure the presence of a carrier "vector" suitable for gene transferred in order to carry the gene of the object to the donor organism.

Discoveries that paved the way to genetic engineering:

Carrier "vector".

Types of bacteria contain a small chromosome called "Plasmid".

Restriction enzymes DNA, you cut it off at specific sites.

Ligases close the gap left by the restriction enzyme.

Select a succession bases in the DNA Sequencing.

Synthesis of pieces of DNA "Oligonucleotide synthesis" for the purpose of identifying genes within the cell for the purpose of diagnosis of many genetic diseases and led to begin the implementation of the Human Genoma Project.

Using "Probe" in the processes determining the existence of gene and diagnosis and genetic makeup of the individual.

The genetic engineering since its birth in the seventies of this century a lot of fear, it is double-edged sword usable for good or evil and see the use in the prevention of disease or treatment, whether genetic surgery that change the genes with other as well as another gene in the filing of another object to obtain large quantities of secretion this gene for use as a drug for some diseases. After the success of the possibility of transferring genes from one cell to another, there were some concerns, including the following:

The possibility of introducing genes that are synthesized toxic material within the cells of bacteria and make them so harmful effect.

The introduction of parts of DNA Tumor Virus in another virus bacteria.

Disable the genetic diversity, where the plants or animals that were subject to genetic engineering are usually homogeneous, making it vulnerable to bacterial and viral diseases and others.

Scientists played down such fears, the development of standards and controls to reduce the risk of manipulation of genes and these terms:

The issue of genetic engineering, genetic since its birth in the seventies of the last century a lot of fear, they double- edged sword usable for good or evil and see the use in the prevention of disease or treatment, whether surgery, genetic change since intra gene in the cells of the patient and the gene in the filing of another object for large quantities of secretion of this gene for use as a medicine for certain diseases while preventing the use of genetic engineering on sex cells "Germ Cells" for the legitimacy of the dangers.

However, the scientists played down such fears, the development of standards and controls to reduce the risk of manipulation of genes, some of these precautions: Controls for the design of laboratories and security measures to prevent the spread or leakage of bacteria and viruses treatment.

Although the benefits of biotechnology in many areas such as medicine, agriculture, industry and conservation of the environment there is increasing data show that diversity of thought began to intervene in the subjects continued to reduce the time from the jurisdiction of social thought and reason. The evaluation process of biotechnology has not yet started, does not believe that the limited studies on the impact of bad to launch microorganisms genetically engineered has no real value. These studies resulting from the overlap of politics in science and nothing to do with the problems faced by our communities.

Protein map "Proteome"

The term "Proteome" which appeared in 1994 to the total pool of proteins present in each cell type the amount of a hundred trillion in each individual and the total proteins produced by cells of the body during different life stages.

After discovering the human genome, which includes (full content of genes (genes) in the amount of the 34 thousands people only, and not one hundred thousand. I think scientists for a long time) and also all the genes inherent in the cells of the body at the present time highlights the important question, what the protein content of these cells.

The type of each protein has to be known as a result of these cells and what function each protein and then what order of these proteins. Asking this question came after attrition rationale the concept of the genome and its consumption and is not enough to know as responsible for stimulating cells to produce the kinds of protein, but only requires the knowledge of the situation in its entirety in routine cases of disease and natural and in accordance with these questions and answers on the back of proteome.

Proteome contains information more complicated and the secrets of the genome is more dangerous than those found in the genome and extensive knowledge and synthetic for more than a million different types of proteins. The concept of proteome is known later human proteome is doing now by scientists and they hope that these will be the beginning of the main achievements under this project, despite the severe difficulties faced by these scientists, in excess of those related to the human genome.

The analysis (cell proteome), reached some of the researchers in 2000 to build automated device called the molecular scanner "Molecular Scanner", which is carried out by measuring by mass spectrometer, from which tens of thousands of known proteins in a single day, and at the speed of more than ten times what was known before.

These researchers also managed to build a million boosted the analysis of protein per day to build bigger infrastructure database proteomics mankind. The draft of human proteome or other whereupon many of the laboratories and big budgets and international companies such as the famous, hybrgenics, Clera Genomio, different research directions, the analysis of three- dimensional protein structure and interactions between proteins, which performed many of the key characteristics of the human proteome would pay off represented by the following:

Specification of fungus or yeast proteome with a single- cell, the first that has been done in the world of proteome.

This project change from how to design drugs in the near future.

The appearance of the so- called science and technology human proteome, which will focus on the conversion of most of the drugs manufactured by genetic engineering and biotechnology.

Chapter Five

Development of the biotechnology

Preface

Iraq occupies a very diversified geographic and climate zone and assumes a great deal of importance. However, its wealth is under U.N. resolutions. Its health and agricultural problems are diverse.

For almost twenty years (1970-1990), its Biotechnological activities have sought to bear on various problems on Agriculture and Medicine. In recent years (1990s), these activities have turned in a very limited way, to use some of new developments in Basic Biotechnology such as Molecular Biology and Genetic Engineering.

Various Centers and Departments are involved in the National Biotechnology research systems in Iraq. The Biological and Agricultural research Centers also integrate such activities into their research program.

Presently, these centers established different types of programs (long term, short term) in Agriculture and Medicine, for generation, development and transfer of Biotechnology. Major components of these programs; include development of technology transfer facilities and establishment of international collaboration in the field of Biotechnology.

Medical Biotechnology in Iraq, as a whole, demands strong collaboration in order to reduce the present gap between Iraq and the developed countries. Whereas, Agricultural Biotechnology seeks to increase food crop resistance to insects and diseases, utilize a wide varieties of techniques, including Biological control and various forms of Biotechnology, such as tissue cultures, molecular genetics and genetic engineering.

This part of the analyses of the major activities of Biotechnology in health, Agriculture and Basic Science in Iraq offers general background information on the development of Biotechnology. Then examines the current of Biotechnology research and potential, and suggests an outline of the strategies for the future of Biotechnology in Iraq.

Iraq, which has already entered the biotechnology era, whether on purpose or by chance, will in any case inevitably suffer the consequences of the widespread use of such techniques in industrialized countries. Therefore, the question is no longer whether such a move is desirable. but more how can best be put into use.

Currently people of Iraq are suffering from poor quality of life because of U.N. sanctions, which resulted in:

- Food shortage and malnutrition.
- Unsafe drinking water.
- Improper sanitation system.
- Poor health care.

This is why Iraq needs to have at least a basic level of skills to make it possible to define and implement policies in Biotechnology and the other technologies. Biotechnology alone will not feed Iraqis or give them better health, but it would be irresponsible not to see it as one of the available tools. A certain number of social, cultural or institutional conditions also have to be satisfied before technical success can be converted into economic and social progress.

This part analyses the major activities of biotechnology in Iraq; then examines the prospects of strengthening biotechnology in conjunction with conventional technologies; and finally it outlines the strategies for the promotion of biotechnology in Iraq.

Historical Aspects of Biotechnology in Iraq

Biotechnological activities were known to ancient Iraqis in Samaria and Babylon. For instance, fermentation processes and perfume extraction were very well-known to them.

It has been reported that the first experiment in scientific research was performed in the palace of one of the kings of Babylon. The experiment was carried out by two palace ladies who prepared perfumes by distillation from plants (steam distillation).

Cheese making is believed to have started in Mesopotamia at approximately 8000 years ago.

In the early 1970s scientific infrastructure construction began and several research activities were planned and set up.

Most of these biotechnological activities were limited to traditional methods to serve their needs; i.e. industrial fermentation, soil microbiology and bioconversion of waste products.

The government showed its interest in providing support to biotechnology by offering to host the first Arab Conference on Genetic Engineering in 1984. As a consequence of the meeting, the council of Scientific Research established "Genetic Engineering Center", which became responsible for all research and development in this field; then in the agreement that Iraq established an affiliation of the International Center of Genetic Engineering (ICGEB) with the Scientific Research Council, as the principle Liaison Institute.

Since then during early 1990s research in agricultural biotechnology dealing with production of Biocides and Biofertilizer and Tissue Culture Technique were carried out by (IPA) Center.

The Ministry of Higher Education in the middle of 1990s, established a new centers for Genetic Engineering, involved in research on biotechnology. The main research objective on biotechnology of these centers involved the application of basic genetic engineering particularly in the following fields:

- Molecular biology: for enzyme production, industrial and clinical diagnosis.
- Cell biology: for animal cytogenesis (chromosome and gene mapping).
- Microbial genetics: for the production of useful microbial compounds through genetically improved strains.

Iraq through a new council of biotechnology at the University of Al-Nahrain (previously Saddam University) is monitoring international development in biotechnology in order to apply some of these techniques. As a result, this council has developed plans for biotechnology and genetic engineering in the sectors of health, agriculture and basic biotechnology. The major components of this plan are:

- Generation, transfer and development of biotechnology.
- Development of trained personnel.
- Strengthening of research and technology transfer facilities.
- Establishment of international and regional collaboration in the field of biotechnology.
-

Approaches of biotechnology in Iraq

Most biotechnological activities, which are applied in Iraq, are limited to traditional methods and serve their needs through:
- Fermentation
- Antibiotic industry
- Single cell protein
- Plant biotechnology
    - Tissue culture
    - Soil fertility, through biological activities
    - Increasing food production through plant cell-culture
    - Bioconversion of waste for food and feed ingredients

Fermentation

Iraq in early 1970s established traditional fermentation industries, bakers yeast production, ethanol, acetic acid, acetone, butanol and citric acid production. In 1970, a factory for making bakers yeast from sugar-beet molasses was established with plans for the production of compressed yeast.

This industry was faced by many problems, especially in dried yeast production:

- There are abundant sources of raw materials for fermentation in Iraq. Large quantities of hydrocarbons, and carbohydrate by-products (molasses) and lignocelluloses waste are found.

- Fermentation of food crops like dates, which enjoy comparatively large market sizes, does not receive sufficient attention. Local research on bioreactors is therefore needed to support the development of new processes and improve the performance of food industry in Iraq. Bioreactors also play a key role in the production of enzymes used in the beverage, detergent and leather industries.

Antibiotics industry

The antibiotics industry in Iraq started in 1970 for the production of penicillin and tetracycline. On the other hand, tetracycline production continued until 1980. This industry was discontinued for economical and technical reasons as a result of sanctions. The involvements of research and development programs have positive effect for restarting bio-industry and production of different types of antibiotics.

As the pharmaceutical industry in Iraq is directed to satisfy the local market by satisfying the market needs, it is therefore necessary to develop appropriate biotechnologies against various diseases that are endemic.

Single-Cell protein

A research and development program for single-cell protein (SCP) production at pilot plant level started in Iraq in 1982. The production of SCP from methanol using local and imported strains of Candida Utilis was investigated. The research plan included an economic feasibility study and the assessment of technological and nutritional aspects of SCP under local conditions. Then, another pilot plant was established to utilize date syrup for the production of bakers yeast and SCP.

The most important achievements of this program were the establishment of pilot research facilities, training of personnel, the nutritional assessment of available commercial SCP products and the isolation of several methanol-utilizing bacterial cultures.

Plant biotechnology

Plant biotechnology applications in Iraq include soil fertility (nitrogen fixation) using yeast strains in a mixed culture and cell conversion. Scientists have carried out research on nitrogen fixation by grain legumes. The main objective of this research was to increase the yields of grain legumes while decreasing the input of inorganic nitrogenous fertilizers

Tissue culture

- In view of the importance of the date-palms in Iraq, a tissue culture laboratory was established in 1979 at the Agriculture and Water Resources Research Center in Baghdad. Another laboratory was initiated at the Genetic Research and Biotechnology Scientific Research Center for the improvement of plant production. Tissue culture laboratories were also established at the Universities of Basra and Musol and also at the Ministry of Agriculture. The latter worked in collaboration with FAQ through regional center for date-palms in the near east and North Africa. The major objective was the commercial propagation of plants by in vitro techniques.
- Then in 1982, the vegetative micro propagation through tissue culture was carried out as a promising technique. However, future research is needed to early following and lack of uniformity of the closed plant.
- Date palm propagation by tissue culture was also implemented at research institutes and universities in Baghdad and Basra. Other species such as lettuce and potatoes were propagated by tissue culture at the department of Biology of Mosul University.

Other activities

- Research at the Faculty of Agriculture and Biology in the Nuclear Research Center was concentrated on the study of a sexual embryogenesis techniques and the induction of mutations by mutagenic agents.

- Agricultural and forest residues in Iraq are considered renewable resources that can be utilized by bio-technological means for the production of food, fertilizers and fuels.

The situation of Biotechnology in Iraq, since the imposition of sanctions

As a result of sanctions, many fields in public health system are affected sector by sector including critical areas such as: food security and nutrition, water resources, women's health, children health, national health emergencies, hospital care, humanitarians' donations and international cooperation. Scientific fields are also affected such as Oncology, Cardiology, Nephrology, Endocrinology, Ophthalmology, Diagnostic Testing and Protection of Blood Supply and Scientific Information and medical education, pharmaceutical and biotechnology inputs.

In agriculture, the U.N. sanctions ban the importing of fertilizers. Shortages in production of crops led to the deterioration in the Iraq's populations and nutritional intake (daily caloric intake).

- Iraq is facing the following three main threats: food supply, health improvements and environmental protection. Millions of people in Iraq under poor and risky growing conditions are suffering from poverty and poor health. They go to bed hungry due to U.N. resolution and security problems.

- Food security in Iraq is unique; it is not related directly to biotechnology problems. Genetically altered seeds are not necessarily needed to feed Iraq. This view rests on two critical assumptions, which we question: The First is that poverty is not due to a gap between food production and growth of population. The Second is that biotechnology is not the only or best way to increase agriculture production.

- Iraq is not able to undertake effective agricultural biotechnology research for its own urgent needs without the scientific support of developed countries. Iraq needs more investment in developing appropriate agricultural biotechnology.

- Iraq which is subtropical country is basically an agricultural country. Approximately 80% of the total area is devoted to cereal production mainly wheat, barley, rice and corn. The remaining 20% includes a wide variety of crops, date-palm, citrus, tobacco, cotton and others. In general the average production level of all the important crops in Iraq is very low compared to those in developed countries. In the last 10 years, plans have been organized by the Iraqi government with cooperation of agricultural scientists to improve both the quality and quantity of crops per capita, using proper machinery equipments, fertilizers and suitable control measures.

Applications of biotechnology in Iraq in the experimental stage

Animal production

The use of biotechnology in animal production in Iraq has occurred, in the field of reproduction, animal health, feeding, nutrition, growth and production.

- In the field of reproduction, new biotechnologies such as embryo transfers, in vitro fertilization, cloning, and sex determination of embryos have been studied experimentally for different types of livestock at the faculty of Agriculture and Biology in the Nuclear Research Center.
- Animal health can be improved with new biotechnology methods at experimental stage of diagnosis, prevention and control of animal diseases.
- Biotechnology in animal nutrition concentrates on improvement of feed; enzymatic treatment, the decreasing of the anti-nutritional factors in certain plants, such as legumes, which are used as feed.

Plant production

At present, more traditional aspects of biotechnology such as the followings are used:

- Tissue culture

The application of tissue culture does not require very expensive equipment. This technology was applied in Iraq to improve local varieties of food, crops for example using traditional methods for propagating potatoes for example.

- Pest and weeds

In Iraq, most of the land is affected by many weeds causing big losses in the agricultural crops. Since 1970s several herbicides were used to control the weeds of corn, cotton and vegetable fields.

Bio-control programs used for controlling pests

The plant production research center is a State Board for agricultural research. College of Agriculture, Baghdad and Mosul Universities, Agriculture and Biology Research Centers and Iraq Atomic Organization are the main research centers in Iraq, which carried out different research studies on the biology, taxonomy and control of the pests. These research centers successfully adopted control measures on the most important pests attacking crops, vegetables and by applying chemical and agricultural methods. Promising results were obtained from many plant extracts against pests, by inhibition pest life cycle.

In Iraq, most of the agricultural land subjected to grow many of weeds causing big losses to the agricultural crops, as many research workers confirmed the positive results of the weed control measures to increase the yield of the crops.

Researchers started to evaluate herbicides since 1965, and mid of seventies, herbicides were applied to control the weeds in wheat, rice, corn, cotton, potatoes and tomatoes. Increased support is needed to expand research designed to develop new herbicides that are not likely to pollute ground water and that will provide reliable control.

Several centers are involved in bio-control programs such as:

- Agricultural and Biological Research Center (Baghdad).
- State Board for Agricultural Research / Ministry of Agriculture Baghdad.
- College of Agriculture / Baghdad and Mosul Universities.

Commercially, research centers introduced two bio-control mutant fungi. Both fungi successfully were applied to control plant parasitic nematodes and soil born fungi on vegetables and citrus. Also the center used another fungus against date-palm stern borer insects.

State Board for agricultural research successfully adopted several control measures on the most important pests attacking field crops, vegetable and fruit trees by applying chemical, biological and agricultural methods. Many insect growth regulators, bio-control agents, fungi and plant extracts were experimentally applied on small and large scale fields. Most of the research studies of the graduate students in the department of plant protection concentrated on the biological control, and plant extracts.

At the present time, the U.N. sanctions which have been imposed on Iraq since 1990, destroyed most of biological control programs and Iraq is facing lack of well trained personnel and shortage in facilities. As a result the first generation biotechnologies used in Iraq such as insect resistance, herbicides resistance are not easy to address any more.

Health biotechnology

- Several centers are interested in carrying on research on health biotechnology. The following may be mentioned.
  - Previously Saddam Center for Cancer and Medical genetics Research (SCCMGR).
  - Institute of Biotechnology and Genetic Engineering for Graduate Students, University of Baghdad.
  - Department of Genetic Engineering, College of Science, University of Baghdad.
  - Genetic Engineering Departments in a number of Universities.
- SCCMGR is engaged in several lines of biotechnological activities in health:
  - Cloning of tetanus toxic gene into tumor cells.
  - Preparation of tumor cell lines in vitro for gene therapy technique
  - Preparation of restriction enzymes vectors for gene therapy.
  - Studies on disorders of mitochondrial DNA in muscular dystrophy.
- Several biochemists have participated in various research projects that deal with diagnosis, and monitoring of several types of tumors.

Future biotechnology

Development of biotechnology applied to food and health to face basic human needs in Iraq

The need for the application of biotechnology to face the basic needs regarding food and health in Iraq is real. There are different approaches such as the development of plant biotechnology, biotechnology applied to livestock production and biotechnology applied to food processing.

So, the suggested components of biotechnology plan include:
- Micro propagation: through e.g. tissue culture for multiplication.
- Genomics: the molecular characterization of all species.
- Bioinformatics: the assembly of data from genomics analysis into accessible forms.
- Diagnostics: the molecular characterization and identification of pathogens.
- Molecular breeding: the identification and evaluation of desirable traits in breeding programs with the use of marker assisted selection
- Transformation: the introduction of single genes conferring potentially useful traits into crops, livestock, fish and tree species.
- Vaccine technology: use of modern immunology to develop recombination DNA vaccine for control of lethal diseases.

Plans for future biotechnology research should be formulated through several priorities:

- Food security.
- Increase and improvement of agricultural production. Breeding for higher- yielding plant varieties and improve nutritive development values. Pest and pathogen-resistant genotypes and conservation of plant genetic diversity.
- Production of pharmaceuticals for the extraction of biologically active plant substances.
- Immunology: Production of vaccines and monoclonal antibodies.

- Use and recycle of agricultural products for the production of ethanol, acetone, butanol and methanol.

Food security

The applications in agricultural biotechnology in Iraq have the promise of bringing about the much-needed requirements in agricultural production, such as carrying resistance / tolerance to a biotic stresses (drought and salinity) and to provide options for better rotation to conserve natural resources.

Iraq is neither in the process of testing genetically enhanced products in number of crops, nor in the process of testing commercial products in the market to-date. Many technological advances are not visible in the farmers fields in Iraq but in the future are expected to provide ways to improve crops in a precise and fast manner. Use of functional genomics to address complex traits, marker assisted breeding to ensure presence of key genes, improving nutritional quality and managing natural resources better by use of efficient monitoring tools, Iraq must be an active participant in this area so that specific needs of food security are achieved.

Agricultural production

Despite the importance of health and industry sectors, the suggested priorities in the plan should have great emphasis on agro biotechnologies for two reasons. First: research on plants for crop improvement directly relates to specific ecological conditions predominant, whereas biotechnology applications for industry and human health are more difficult in Iraq. Second: preliminary data indicate that most biotechnology research activities in Iraq relate to agriculture.

The following classes of agricultural biotechnology are suggested to be used in Iraq in the future:

- Gene transfer technologies, which provide transgenic plant, resistant to many pests' pathogens, herbicide resistant to stress such as temperature, drought and salinities.

- Non transgenic biotechnological approaches for improving the efficiencies and effectiveness of conventional plant breeding methods.
- Technologies for better monitoring of natural resources and environment.

Additional suggestions for the implementation of the plant are the following:

Plant biotechnology

- The establishment of biological treatment plants for sanitary wastewater and utilization of the treated wastewater for landscaping and agriculture.
- The establishment of the commercial production of inoculants such as Rhizobium and developed efficient methods for recycling agricultural waste such as beet molasses and maize and rice straw. Research should be carried out on the use of bio-fertilizers to increase rice yields in Iraq.
- The use of biotechnological techniques for the development and improvement of bio-insecticide for control of plant pests.
- Increasing protein content of rice by the application of biotechnological techniques. Rice is one of top five cereals grown in Iraq with its unique capabilities desirability to grow in stress conditions it is a crop of choice for Iraqi people. Rice has received comparatively less attention for research in general.
- The creation date-palm clones resistant to disease, and the application of tissue culture to improve date-palm varieties.
- Future research is needed to overcome the difficulties related to early flowering and lack of uniformity of cloned plants.
- Production of secondary metabolites, by tissue culture, the selection of plant cell lines for stress tolerance to salinity and drought and also the production of virus- free potatoes planting material; and the micro propagation of plant.

Animal biotechnology

Biotechnological application in livestock and fish production, and the adoption of embryo culture to improve local animal breeds through embryo transfer technology, are samples on pre-implantation and embryo freezing.

Microbial biotechnology

- Microbial biotechnology for ethanol production from sugar by-products and methanol production from Agro- industrial wastes.
- Microbial genetics: elimination or degradation of pollutants transformation of cellulolytic nitrogen fixers; construction of Saccharmyces cervical strains capable of cellulose, cellobiose or lactose consumption.
- Proper technology to convert biomass into bio-fuel and biogas to convert agricultural biomass and animal droppings into bio-fuel and manure; of the various wastes used for biomass production. Rice straw is one of the possible applications.
- Bioconversion of lignocelluloses wastes to protein- enriched fermented materials followed by the production of microbial biomass from the hydrolyzed cellulose product.
- The use of bacterial treatment for the removal of oil and chromium; and also in the nitrification-denitrification process to remove ammonia.

Health biotechnology

Medical Biotechnology in Iraq, as a whole, will demand a strong collaboration in order to reduce present gap between developed countries and Iraq to achieve this aim:

• A center of bone marrow transplant should be established in one of the hospitals. This requires the availability of experts, equipment and materials.
• Iraq is facing cancer and genetic diseases hence, research projects in gene therapy should be initiated.

Pharmaceutical industry

The pharmaceutical industry in Iraq should be expanded and developed so as to meet at least the local requirements. Biotechnological techniques should be introduced.

Environment

- Applications of natural occurring organisms (e.g. yeast, fungi and plants) should be used to convert hazardous substances in soil.
- Using microorganisms' pollutants from sewage systems to clean up industrial sites.
- The use of biotechnology to avoid pollution is of increasing importance such as the use of bioreactors to treat hazardous products.

Forensics

Applications of technologies for forensic science.

Bioinformatics

This new discipline (biology and computing), which will be the core of biology in the 21st century, should be used for measuring and monitoring thousands of genes at one time. This computer-aided bioinformatics will stimulate future developments in the pharmaceutical industries.

Cooperation with International Agencies

- Cooperation with Islamic and International agencies and countries are required.
- Well trained scientists from Arab and Islamic countries, directly involved in the training and transfer of various biotechnologies are also required.

- Post-graduate short training courses sponsored by international organizations such as the United Nations Educational, Scientific and Cultural Organization (UNESCO) should be organized in the various fields of biotechnology.
- Training of medical personnel in bone marrow transplantation, to help in gene therapy especially for cases of leukemia and lymphoma.
- Broadening the biotechnology base is a must for characterization, collection and conservation of germplasm that is already in the gene bank collection around the world and providing information for collection of gene pools that are not currently available in gene bank.
- Strengthening capabilities, developing projects, visits, and training programs of mutual interest to all participating countries in the following areas of biological control:

  - Exchange of biotic agents on a case to case basis.
  - Mass production of host insects and natural enemies.
  - Biological suppression of crop pests by developing joint projects.
  - Computerization of information and networking research organization in different countries.
  - Training in different aspects of biological control.

Establishment of Islamic Biotechnology Center

On the light of what has been presented in this paper I believe that, one would come to the conclusion, that an Islamic Biotechnological Center is really needed. Islamic countries which have longer experience in biotechnology could greatly help in establishing such a center.

Ethical issues related to biotechnology

The Islamic world needs to have sharp opinions on various current issues related to biotechnology and genetic engineering such as genomics, human cloning and genetically modified organisms.

The National Programme for the Biotechnology

Bio-technologies represent the modern mosaic of knowledge, rooted in the world of genetics, biology, evolutionary biology and molecular knowledge of chemistry. Bio-technologies are unique among other technologies for being a tool for dealing with life itself. They include technical applications that use the vital systems of organisms or their components or products to modify products or vital processes for specific purposes, and therefore include many operations and have wide application in agriculture, industry and other sectors.

Despite the benefits of bio-technologies in the fields of medicine, agriculture, industry, however other biological fields increasingly intervene with many other subjects and indicate that the diversity of thought began to appear. Therefore, a significant problem surfaced when preparing the National Programme for bio-technologies. Then, there was a significant overlap between the main axes of the programme and its affiliates. It developed themes and sub- themes according to different areas of bio-technologies that can be investigated by the Iraq meaningful progress particularly in the areas of food and health and strengthening pharmaceutical industries.

Major components of the programme

- The biotechnology and food security
- The production of bio-pesticides.
- The production of bio-fertilizers.
- Tissue culture.
- The production of potato tubers and seedlings free of viral disease
- Production of broad Date Palm
- The production of Fruit assets.
- Production of secondary materials

- Education programs and improvement.
- Technical embryo implantation in cattle.
- Manufacture of food products with curative nature.
- The production of liquid sugar from Iraqi dates.
- A study of pesticide residues in food.
- Improving the nutritional values of feed.
- Single-cell protein production.
- Improving the conditions of storage of foodstuffs.

Health and medical technologies

- Gene therapy and diseases
- Markers of neoplasm
- Genetical markers
- The production of vaccines
- The production of antibiotics
- The production of hormones and enzymes
- Vaccine production
- The production of antibodies to microorganism and toxins from snake bites and insect bites
- The production of antibodies (monoclonal antibodies)
- Development or establishment of cancer lines in the laboratory
- Transfer of bone marrow and cultivation of bone marrow
- Using molecular indicators in human lymphatic cells
- Diseases of hereditary Cancer
- Diagnostic kits
- Forensic-genetic fingerprint

Biotechnology and safety pharmaceuticals

- Extraction of drugs from living organisms (plants, animals and microorganisms) for use against cancer and other diseases.
- Extracting of active substances.
- The production of pharmaceutical materials and medicines through the revival of genetically modified microorganisms.

- Preparation of medicines from medicinal herbs.

Plant Bio-technology

- Plant tissue culture, both at the level of cell or tissue or at the level of protoplast.
- Improving the quality attributes of different crops.
- Production of plants of potential environmental conditions.
- Production of plants resistant to pests and agricultural bush.
- Increased production of medical and pharmaceutical plants and medicinal herbs.
- The genetic diversity of plant.
- The production of fertilizers and bio- pesticides.
- Development of insecticides to control plant pests.
- Reproduction development of resistants to diseases and the application of tissue culture.
- The production of secondary materials using tissue culture.
- Production of plants that have the capacity to fix Nitrogen.
- The production of new crops resistant to pesticides and pests salinity.

Animals and microorganisms technologies

- The production of genetically modified animals characterized by the attributes of high productivity.
- Animal tissue culture.
- Improvement animal resistance to environmental conditions.
- The production of veterinary vaccines.
- Artificial insemination.
- The diagnosis of hereditary diseases and germ using the PCR (Polymerase Chain Reaction).
- The cultivation of embryos.
- Applications of biotechnology in improving livestock.
- DNA technique.
- Vaccination of livestock against diseases.

- Transmission of embryos in cattle and fish.

Microbial Bio- technologies

- The production of hormones
- Fermentation
- Producing bacterial strains of anti- cancer
- The production of bacterial strains of high productivity of enzymes
- The production of bacterial strains of high productivity of antibiotics
- The production of microorganisms strains which have desirable qualities for use in bio- pesticides and fertilizers
- Producing microorganisms strains with high productivity of industrial materials
- Ethanol production by microbial Bio-technologies from sugar and methanol production of industrial and agricultural waste
- The removal and crushing and transformation of nitrogenous cellulosic fixtures
- Bio-conversion to cellulose waste materials rich in nitrogen

Environmental Bio- technologies

- The Use of living organisms in purification of heavy metals of the environment.
- Preparation biocides to combat agricultural pests.
- Preparation of bio-fertilizer to improve agricultural product.
- Finding microorganisms to revive the disintegration of some hydrocarbons and turned into a simple compound.
- Bio-technologies in the ecological balance.
- Bio-treatment of sewage.
- The use of microbial treatment to remove oil, chromium and ammonia.
- Using natural organisms to convert hazardous substances in the oil.
- Clearing industrial sites and avoid pollution.
- Oil Pollution by breaking chemical compounds.
- Isolation of bacterial strains that have the ability to remove sulfur.

- Using Bio-Markers and Bio-controls for detecting pollution levels.

Bio-technologies of Water

- Aquaculture techniques
- Techniques of bio- actors of water
- The use of bio- sensitivity
- The use of drugs and vaccines to preserve the health of fish.
- The cultivation of fish.

Biotechnology and basic sciences

- Genetic engineering techniques
- Genetic fingerprint (D.N.A. Finger Printing)
- Production of restriction Enzymes
- Production of standard D.N.A.
- Diagnosing the production of certain genes
- Amplification of genetic materials

Genetically modified organisms

- Genetic manipulation of farm animals to produce therapeutic human proteins
- Methods of genetic manipulation of farm animals
- Genetically altering poultry for the production of therapeutic proteins
- The use of animals in the production of genetically therapeutic proteins
- Adverse effects on animals that may arise due to genetic manipulation
- Genetically engineered animal, models for human diseases
- Genetically modified plants
- The production of much meat with low fat
- Improving the quality of protein from milk cows

- Gene transfer techniques to provide resistance to many of the insects and plants resistance to high temperatures and drought salinity

Bio-technologies in industry

- The production of antibiotics
- The production of enzymes and drugs
- Production of energy materials such as ethanol and methanol and acetone
- The design and analysis of Bio- reactors

Transfer Bio-technologies

- Technique of PCR (Polymerase Chain Reaction)
- Technique of PCR- STP (PCR with Short Tandem Repeats)
- Other techniques

Cloning

- Cloning techniques

Genome (genetic content)

- Studying the genetic content of the bacterial isolates
- Isolating and purifying D. N. A.
- Determining the content of the Plasmid isolates
- The safety of animal products and genetically modified organisms
- Building gene maps of plants of economic importance

Studies related to bio-technologies

- The systems and regulations for genetically modified animals
- The importation of transgenic animals

- Economic feasibility of genetic modification in animals farm to produce therapeutic human proteins
- Controls systems in the use of genetically modified organisms
- Ethics and social values and bio- technologies
- Bio- informatics
- The Bio-safety and bio-technologies
- Laws regulating the use of genetically modified crops
- Medical, religious and security considerations of bio- technologies

Chapter six
Technology of Education and higher education

Contents

- **Preface**
- The general features of the education in Iraq
- **Technologies of Education**

.Contemporary technologies in education

.The challenges of Iraqi education

.Technologies of Scientific knowledge . Brain Drain and education

. **Global Experiences in Education**

.**Technology and Higher Education**

Preface

The education system in Iraq, including a variety of setbacks before and after the U.S. occupation, characterized by several more effective than other service sector and it can be argued that this system had been addressed either by mismanagement and neglected it infrastructure (school buildings) or in some cases the supplies, means and that the income of the teaching staff and educators. The motivation to work was decreased, resulted in children dropping out at all academic levels in the largest primary school. The number of students per class was increased the number of schools that are running in the school building become one tutoring phenomenon that hit the school administration at all levels as one type of corruption.

However, according to UNESCO reports, Iraq in the period before the first Gulf War was one of the best educational system in the region. The rate of enrollment in primary education amounted to approximately to 100% and that a good percentage were able to read and write, 55% of males and females 23%.

During the review of the previous note, that the political circumstances that have passed on Iraq has affected the education system accounts largely, spending too little on education, shows the government's neglect of education, caused the educational system to find low, if any, of the system "The weakening of the material capabilities and the lack of qualified manpower cause the decline of the educational system". The financial and natural resources were not used, as well the economic side has a great impact on educational systems, and that education was not associated with the development- factors. There is no human of manpower capital plan and production of a humanitarian and investment. As for the social aspect of the Iraqi family that was very conservative, resistant to change, and the of attention to education, especially female education system that has also been affected by divisions of Iraqi tribal and clan differences in addition to ignorance, poverty and disease.

The infrastructure for education of three provinces in northern Iraq with less damage, where UNESCO, UNICEF carried out the implementation of education teaching programs were jointly witnessed a considerable development of educational institutions that has also been providing educational materials at all levels of education and increased the ability of individuals to access to education. There has been development of educational facilities of the elements to spend on the process of construction and purchase locally. While the supply of the equipment, through the United Nations under the oil for food program was too low.

Dr. Ala Alwan indicate that the documented information on the budget of education in Iraq are not available while the reports of the United Nations and World Bank, stated the budget of basic education before the nineties amounted to 680 million Iraqi dinars, 50% of primary for education and 27% for secondary education, and 20% university education. During the nineties, especially 1997 the education budget through the oil- for- food role of 700 million in the form of goods, equipment and projects. To return to the pre-war period the education teaching budget represents 5,2% of gross total national incomes in 1970 and 4,1% in 1980 and stood at the end of the Iran-Iraq war in 1988-1989, 6% of gross total national income.

According to the statistics for the year 2001-2002, the enrollment of about 54000 children (4-5 years old) have been registered. The Arab Human Development Report issued by the United Nations Development Program, in 2003, indicated that the enrollment rates, "net enrollment ratio" in the pre-primary education was 5,8%.

Education is compulsory for a period of 6 years, which works to enable all the children of Iraq from the end of their six years of age to develop their personalities aspects of physical, intellectual, spiritual and become citizens of sound mind and body, creativity. The primary school education in primary schools was free of charge and the Ministry of Education provides students with textbooks and educational materials free of charge. The report published by UNESCO in 2003, include the proportion of the total enrollment of "Gross" enrollment ratio was 107.8% during 1990 to 1991 and dropped to 98% in the year 2001-2002, only 72.8% of children who entered school in first grade, continued until the fifth grade in the year 2001-2002, compared to 75,6% in 1990-1991.

The general features of the education system in Iraq

Iraq's education (teaching) is compulsory until the completion of primary and the Arabic is the official language in addition to teaching the Kurdish language in the Kurdish areas.

For example, the role of culture in education is curricula where the curriculum needs to pot culture in- depth thought, understanding perception, intelligence strength of the presence of intellectual inheritance, scientific and humanitarian law.

The cultural role of the teacher is supposed to be broad literally that does not restrict the methodology, knowledge, cognition, mixing of culture and knowledge should be fullified by the nature of the subject that the characters of the teacher are required for the cultural imperative for social and scientific considerations.

The components of the Iraqi people and factions, all as the result of centuries of brotherly relations between these components. The sharing of Arabs, Kurds and Turkmen, Muslims and non Muslims live in sweetness of hope and all burned by the fire of oppression and tyranny. In the fields of the national struggle is the blood of Iraqis, in the struggle against the enemies in the field of culture, thought and creativity that reflect this diversity but this unit in the works of writers, creative people thinkers in the arts have all been educated by Iraqi culture certain thought, literature enrichment and wonderful art old and new.

In our contemporary culture many names in the field of political thought, and literature, the area of musical art, the area of architecture. The Turkmen are not much less contributing to the civilization of the Iraqi people in the field of poetry and literature.

Technologies of Education

Globalization is a group of political, cultural and economical operations that penetrate the borders of a single country, leading to the convergence of parts of the world, and leading to differing views of intellectuals on the concept and objectives, and societies differ on the extent of assimilation, acceptance or rejection. A proponent of globalization refers to a form of streamlining relations between the nations of the world. As opposed to globalization, it is believed as a balance between actors, the strongest is at the level of global capital that dominates large companies, and also it rejects the view that globalization is a kind of domination by powerful countries on the vulnerable, where consumers have been the last of knowledge and thus globalization have negative impacts on economy, education and politics.

In the face of globalization it requires developing countries to revert, then to revitalize the relationship between education and economy, development, culture. Then education and economics of future generations are the basis for developmental spirits of the knowledge society and culture.

Education instructions produce economic public good, where academics enjoy a level of economic and working conditions of eminent there is a difficulty in governing the university by the market, but the reality of education in developing countries, including Iraq's education structures in its content and methods is traditional but suffers from problems of administration, organization and funding. Therefore, the first step of confrontation to operate is to understand the times and challenges. However, the positive effects of globalization on higher education have emerged by increasing numbers of education types linked to the internet that offer its services globally.

Globalization is vulnerable to education making reform of this education and the need to make it more important than ever, and that any neglect of this sector threatens the entire development process. The pressure of globalization makes it necessary to allocate adequate resources for education section therefore there should be reform of this sector at the enterprise level and the system as a whole.

The education is helping developing countries. It benefited from globalization through technological developments, mostly in developed countries and education could be a fundamental tool because it helps developing countries to reap the benefits of globalization, the latter can help countries in attracting foreign investment, and help any country products to benefit from education. Finally linking the globalization to education provides opportunities for improving living standards, India for example, benefited from globalization by building the computer industry.

Some believe that education are international outlook and global trends in nature and not on globalization trends, where the globalization of education viewed as commodity business governed by market forces. The positive aspects of globalization from the perspective of increasing access to education, and reducing knowledge gap in developing countries, at the same time, negative aspects of globalization threaten the education of developing nations. The globalization of the economy includes the agenda of education business for the International Trade Organization "WTO". This is not for the contribution of education in development, but as services or commodity trade. Education has become a market that generates substantial funds, as well as the globalization has affected extensively resulted from the policies of globalization, low support of the public sector in a free-market economics, led to a commercial-run education. Several European educations, which depend on market cooperative economics, did not follow the way Anglo-American universities in making managed trade.

Contemporary technologies in education

Technology is comprehensive, targeted process and systematic application of scientific knowledge in education, play multiple roles in education that have been directly under the concrete aspect which are used in everyday life and the intangible processes, systems, complex tasks that should be implemented, planning, management and evaluation.

The role of technology education in educational reform, in which would indicate a role to play in the reform projects or educational innovation such as curriculum development and design of actual systems and are now in the U.S. educational system a key tool in the movement of educational reform, which includes the educational designer to contribute in three variables.

According to international development indicators from the World Bank 2003 the number of computers in education in Egypt was in 2001, 48816. By way of comparison, we find that number in Malaysia has reached 121,859. The population we find that the number be in Egypt 75 computer tutorial per hundred thousand inhabitants, while the total number in Malaysia, 512, or about seven times that of Egypt.

Higher education real crisis in the Arab countries lies primarily in the foggy relationship with the labor market, which fall the performance and productivity of higher education institutions.

The gap between them and the challenges of the society was expanded, failed to provide essential services and the doubts were increased about the ability to attract productive sectors. The cost and quality of education have become the most important crises facing the Arab high education institutions.

The turnover at the university and higher education institutions in developed countries imposes significant consequences on the situation of higher education in the Arab countries. The higher education in the Arab countries varies far from its counterpart in developed countries, in substance, as mentioned, followed by strong doubts on the possibility of catching higher education in the Arab Countries and counterpart in the west. On the other hand, higher education in developed countries, which passes in the context of globalization, the rule of the market, the widespread of the higher quality and fundamental shift from its social and guaranteed basic ways to organize, will affect if not have an impact on the Arab Universities.

Therefore, the most prominent challenges that confronted the Arab University, was the achievement of the traditional functions of the university teaching, research and public service (consulting). There are number of channels of higher education in Arab countries reflected the reality of Education, as the "futuristic vision" for higher education in Arab countries.

The report of the development strategy of the Arabic Education in 2003 quotes, " The future of the Arab nation in the near term and long term depends on higher education". As a way to prepare manpower and specialized generation of thought, efforts must be paid on the preparation of researchers and leaders in the areas of employment, production and management of the renewal of culture.

Development of higher education and university in the Arab counties should be carried out through liberalization of its shed, both the government and non-profit societies without abandoning its responsibilities to the state. The establishment of the structure of higher education is characterized by diversity and flexibility, without rigid restrictions. The institutions are subject to constant evaluation and review and establishment of channels for joint action between higher education institutions and state organs.

Upgrading (development) in general application of the knowledge of the educational institutions yields educational systems, materials and equipment. The future of university education in the context of social transformation and strengthening of democracy in education depends largely on the role of Arab universities in the qualitative production, acquisition and resettlement and dissemination of knowledge.

Examination of the reports issued from the Federation of Arab Universities and the Arab Bureau of Education for the Gulf States. Arab Organization for Education, Science and Culture Organization, the United Nations Development Program, UNESCO, shows that the quality assurance testing in the Arab countries.

The challenges of Iraqi education

Iraqi education face serious challenges including challenges associated with internal structure of education and philosophy, methods of science, factors governing the forest and the quality. External challenges are associated with knowledge society and Iraqi education face these challenges, so it necessary to reconsider the philosophy and goals of education and activating the role of Iraqi education in partnership with community. It is also necessary to work on raising the quality of graduates, deepening the Iraqi identity in education, developing a new vision for tertiary education, establishing a strong infrastructure of modern communications network and the introduction of tests.

Iraq is living at the present time in one of the darkest periods throughout history, where it faces serious challenges linked to its ability to survive. They affect its security and stability, and even threaten its existence and entity.

The challenges facing Iraq are many and varied and overlapping, making it vulnerable to military occupation. Great states came to Iraq with pretext of freedom, democracy and human rights, which claim the right to be invalid by the pillaging, looting, and vandalism, and humiliation. If the basis of the challenges facing Iraq is the failure of different dimensions and forms, then the distinctive way to address these challenges is to develop education and leap from the reality of underdevelopment in all its aspects and limitations, to the reality of progress in all its dimensions and prospects.

The education carries the most important responsibilities of many preparations, development; improve performance, and transferring Iraq from an under-developed state to a state in the phase of progress. The education, like the Iraqi society is facing many problems that shackle its movement, and reduce its role in improving the performance of Iraqi rights. Some of the problems faced by Iraq as well as the education.

Technologies of Scientific knowledge

Knowledge is the product of mental process of education and thinking. It is the basis of force and earnings, constitute the most important components including the work and activity, particularly in the relation to economy, culture and education, which is also an economic asset. Furthermore, it includes the human being and his care and preparation the main sources of knowledge.

Other, believe that knowledge is the awareness and understanding of facts or acquire information through experience and means to acquire the unknown and self-development and is directly related to information, education and communication. The knowledge is also the most important component included in any work or activity, especially with regard to economy, society and culture.

The World Bank report entitled "knowledge is the path of development" explains, the knowledge and conditions, the gap between North and South at the end of the twentieth century and that scientists in the world are unevenly distributed between North and south.

Moreover, the knowledge society is characterized as one of the most important product or raw material. The knowledge of modern technology is used in their community and not be complaint to the presence in the same geographical location and it became one of the most important components of capital in the current era which is

supposed to be free; free application for the benefit of community and should remain free.

The transmission of knowledge of the individual is called learning, the process of receiving knowledge through study and educations, while education is the process of making the learner acquire knowledge and skills, university faculty is exercises education and student is learning.

Knowledge society is defined as a range of convergent interests, trying to take advantage of pooling their knowledge in areas that are interested. The knowledge society is of post-modernism linking the knowledge economy, generates profitable commodity. Knowledge society requires the potential and special skills and abilities, super sophisticated infrastructure, natural resources and minds capable of producing knowledge and converting it into super economic progress. The challenges of knowledge society emphasizes the importance of education, and this requires improvements in education systems in order to transform them into the production process of knowledge.

Education is the foundation factor in the relationship with the knowledge society, which facilitates all exchange processes. The advanced stage of acquiring knowledge is the education which is the wider entrance to the knowledge society.

Knowledge society which is called the community of the twenty-first century (third millennium) has the potency for production of knowledge and converting it into profitable commodities which led to the strength of this community and adopt a set of values including:
- Intellectual flexibility.
- Team work.
- Adoption of democratic values.
- Promotion of intellectual diversity.

The infrastructure of "knowledge society" includes:-
- Physical infrastructure.
- Technological infrastructure.

These infrastructures denotes the facilities, services and logistics needed by the community such as, means of transportation and means of communications.

It is worth mentioning that the knowledge society is the basis of the information society, according to the report that was issued by the United Nations, educational, scientific and cultural organization "UNESCO" in 2005 under the title "From the Information Society to a Society of Knowledge".

In Iraq, the knowledge society is defying Iraqi education for being easy to be challenged, since knowledge society is advanced, and Iraqi education is backward. Entering into knowledge society requires providing sophisticated infrastructure of communication technology, favorable climates of stability and contemporary education system with new techniques that emphasize the supreme actual operations.

Production and dissemination of scientific Knowledge

The production of knowledge is an advanced stage that acquires knowledge and can be measured in this production through scientific publications, patents and innovations. As, noted, a significant increase in the movement of scientific publication in the Arab countries, where the number of publications of the Arab scientists elevated annually from 465 bulletin per year during 1976 to about 7000 paper in the year 1995 in different scientific fields such as medicine, agriculture, basic sciences, according to the human developments report for the year 2003. The index of scientific publications, which measured the number of research published internationally, per million persons per year.

It was noted that a wide gap between Iraq and the nations of the world in the production. In order to reduce this gap substantial changes has to be carried out in the education system especially the universities, where there is a need, to build up, a modern communications network, the rules of knowledge, new information, participation in the information systems and disseminating a culture of team work education. The fingertips of the Ministry of Education strides for the insurance of scientific journals in various specialized scientific disciplines and humanity.

The number of workers in the production of knowledge is rising steadily but there are weak response to the demands of the market in higher education systems of Arab countries and shortcomings in the education requirements in the fields of science and culture.

There was a scientific awakening without any producing knowledge such as dissemination of science in which the latest of scientific achievements in the whole world can be found.

The discussion on scientific research publishing is extremely important, because at this stage of the development of science it plays very effectively to the enrichment of scientific research, expanding its scope and providing them with the achievements of researches, as well as to compensate for the lack of specific activities and creative achievements.

Global Scientific Publishing

Scientific publishing is concentrated in industrially and economically advanced countries. Nine countries produce 90% of the global scientific production registered by scientific information institutes. The scientific production of the developing countries 6% of the international scientific production.

It is worth mentioning that India then China are pioneer in production of scientific publication and have maintained a rate of 1:5. Egypt has advanced position in the list, despite the apparent increase in the quantity of scientific production in developing countries.

If we compare the rates of publications of domestic and external level, in different continents, we will find out that researchers of East Asia and Latin America are publishing more locally, whereas in Africa the publications high abroad. In developed countries the publication abroad does not exceed 20% for France, 25% for Japanese, and 12% of the overall European researchers.

All periodicals published in languages other than English are not worthy deserving the attention. The statistical reports that included researches from English speaking countries show that very low percentage published in other languages, while we find that 17% of researchers published in English for French-speaking and 36% of Spanish or Portuguese-speaking.

Scientific publishing in developing countries

Studies that rely on international banks data incline to divide researches of the third world into two categories:

- Those who promote their products abroad in international periodicals with specified weight are taken in consideration.

- The second category is those who submit a local science without great value. The weakness of this provision despite the wide spread and the local science is not necessarily low-quality.

It was evident from other studies, conducted at the East Asia countries, where local scientific journals had reached a level of sophistication and researchers publish with options and not because they insist of inability of external publishing.

The articles of researchers in developing countries received much attention when published with researchers from industrialized nations. This is a very important point that is the choice that must be carried out by researchers from the third world research in science that prevail globally and the trend towards solving local problems, even if it was not expected that this research sheds worldwide attention.

All periodicals issued in the third world do not receive the attention, whereas

- Among 201 chemical periodicals, Brazilian institute of information rely only on six.

- Among the nearly only 200 scientific periodicals published in Thailand, it depends only two; South Korea rely only on one.

Knowledge economy and Technological Incubators

Knowledge economy represents a new type of economy, different from the old economy, which was based on land, labor and capital as factors of production while the new economy adopts the so-called knowledge economy based on factors, including technical knowledge, creativity, intelligence and information.

United Nations estimates accounts 7% of world GDP. Thus, the knowledge economy must be the main engine of knowledge for economic growth. The knowledge economy is characterized by innovation, education, and infrastructure of information technology. The main forces that appose to the knowledge economy are working to change the rules of trade and national capacity in the knowledge economy, including:

- Globalization.
- The information revolution.
- Proliferation of communications network.

Knowledge economy plays an important role in the knowledge society, globalization that is dependent on the economy; it emerged from; which is also contributed to the result of knowledge in technology developments, according to the following requirements:

- Establishments of organizations and new economic rules.
- Opening up world markets.
- Redrawing the map of the world economics.
- The emergence of new centers that depend on world trade.

The emergence of the knowledge economy has led to emphasis on the importance of education as a key to economic success and knowledge society which is closely linked to the knowledge economy. All these are dependent on creative mind generated from higher education, which contributes a significant role in the production of knowledge. Therefore, there is a relation between education and knowledge economy and knowledge economy relies on knowledge production and the production of knowledge is one of the most important functions of modern education.

According to the logical perception of the importance of the knowledge economy and adoption of standards the UNESCO institute for statistics adopts the following indicators:

- Total expenditure.

- **Expenditure intensity.**

- **The link between education and the labor market.**
- **Strengthening the IT infrastructure and knowledge.**
- **The integration of contemporary techniques in the processes of learning and teaching.**

Trying to remove the traditional features of education.

Brain Drain and education

Migration of scientists is due to various reasons, effects some, overlapping in some, and these create psychological climate, related to scientific and incentives conditions. Some of the reasons are social in the homelands, others are materials, may be related to the needs of living.

The phenomenon of brain drain has motives of social, political, and personal nature. The social is characterized by difficulties that faced the developing countries in strengthening the shaken scientific planning in developed countries. One way to keep scientists from migration is to treat the fundamental faults by working to link with national policy, to introduce the idea of scientific planning, providing possibilities for scientific work and atmosphere.

It is an unfortunate facts that the money spent on scientific research and development of the university, in all Arab counties up to 260 million dollars only, while the states of Western Europe during the sixties spent 6 billion dollars per year. United States of America spent 24 billion dollars during the same period. The value of spending on research has increased with the beginning of the eighties to nearly 40 billion dollars. This has encouraged migration of scientists, for example, Iraqis abroad had been attracted to the atmosphere of academic and scientific facilities, methods and the possibility of attending scientific conferences, symposia, magnitude of printed and published by specialized magazines and periodicals.

The phenomenon of brain drain is the most important global problems which recorded at the international level and regional level as stated.

Recent studies indicated that the organization for economic cooperation and development which includes 30 industrialized states, that the immigrant enjoyed a degree of education.

It is worth mentioning that the phrase "Brain Drain" derived by British was used to describe the loss of scientist, engineers and doctors and that UNESCO defines immigration as kind of abnormal types of scientific exchange between the states, as a reverse transfer of technology. The Gulf center for strategic studies in 2004 indicated that western counties had attracted to the west no fewer than 31% of the brain drain for developing countries by about 50% of doctors and 32% of the engineers, and others within the intellectual trends. The west is perceived to the issue of brain drain from standpoint that they reproduce underdevelopment in developing nations.

The risk of brain drain may vary from one state to other, but effects remain similar- in that, (brain depriving in human resources).

The "World Organization for Migration" estimates those developing countries supporting the United States, Europe and South Asia at 500 million dollars annually. On the other hand, the World Bank estimates that one hundred thousand foreigners from industrial nations are working in Africa at a cost of four billion dollars annually. Certain social values prevailed in the traditional farming communities would also decline; especially that migration was characterized as external migration of males. Therefore the migration process may lead to partial destruction of the wealth of mankind.

Brain drain that began after World War II included developing and developed countries spearheaded by Britain, France, Germany, Sweden, Switzerland and Japan. The United States of America is not included, but limited to become a terminal brain drain, from other countries. From a historical stand point, this kind of migration was due to Phoenician and Golden ages of Greek, Roman and Arab civilization.

Theoretically migration and mobility of scientist across the centuries, from, country to country is considered as one of the features of scientific development. UNESCO in 1955 considered brain drain as "the impact of migration from the effects of human solidarity".

It is noteworthy, to study the effects of brain drain on the Arab countries in the perspective of a strategy to develop higher education. The competencies move highly qualified group of individuals from one Arab country to another. This brain drain has not received international attention, only at the end of the sixties and seventies following the transmission of certain competencies to the industrialized countries, where moves brain abroad to more advanced society to increase productivity, but at the same time causing a loss for the country of origin.

It is estimated that Egypt had provided about 60% of immigrates to USA, and Iraq has increased its share significantly after the nineties, followed by Syria, Jordon, Palestine. The UNESCO has chosen Egypt from among those most affected by the brain drain, but did not contribute dramatically to solve this problem

Global Experiences in Education

This system is one of the political viability of contemporary Japanese renaissance; with political guidance the national loyalty strengthen the values of collective beliefs was upholded that support national affiliation. The Japanese educational system played the greatest role in the economic and social progress achieved by Japan since the end of World War II. This system was subjected to ongoing development and amendment. Despite the departure of Japan as a state of war which broken, the Japanese leader have legislated plans and programs and tracks in the field of education and then strengthen the principles of education and then rebuilding of the Japanese citizen.

The development of education and learning in Japan was carried out through the many stages, the antiquity and the stage of European culture and friction of education affected with Buddhism, Christianity, and the reign of Emperor Fiji and developments of education in the twentieth century and the declaration if World War II.

The United States of America as an occupier of Japan have developed a new philosophy characterized by:
- **Prohibition the military educational and instructional activity.**
- **Re- organization of Japanese education in the light of the new philosophy.**

The Ministry of Education in Japan has set goals through the topics of National Education in accordance with an integrated system that participated all instrtuition of society and the interest of leaders in teaching moral education at all stages of education from primary school to university in theory and practice and interest of the events that lead to awareness and concepts.

The Japanese education system used its capacity and strength in spirit and correction and to adopt such stringent qualities of the Japanese nation, which provides models for the people of national service and beyond any individual effort, as well as that Japan has not much affected by the West except for a limited period and then went out to establish a scientific base technical, industrial, professional direction on the preferred to theoretical studies. Finally, we can summarize the reasons of the Japanese experience success according to the following:

- Japan's appreciation of the value of human beings.
- The role of the human element in the distinctive development.
- Reliance on the Japanese language as a key in the various educational institutions.
- Religious heritage has been and remains a distinctive presence of a mixture of religions.

Financing of the Japanese education

The central governments bear the expenses of education with the assistance of regional boards of education and local councils for education and spending on schools in the towns and villages. The central government and regional government bear the salaries of teachers equally, subsidy for regional boards of education, municipal and private schools and research organizations.

The Japanese Experience in educational planning

The educational plan focused on teaching the intensive care unit in basic education and free education up to secondary school and Japan adopted a system of apprenticeship "Industrial Studentship" within the factories. The concern for the individual and the development of abilities and inclinations into account, providing training as part of quality education were increased.

The educational system in the United States of America

That the educational system of U.S. has passed several stages, including:

The first immigrants/ the nineteenth century, the primary education such as reading and writing to be able to pay its expenses and its features during this period:
- Dependence on the European system and the English language are common.
- Introduction of Sunday school and nursery schools of the German character.
- Use a system of compulsory education.

The general objectives of the education system in the United State were:
- Providing the learning for all children and young people.
- The education should be free for all citizens.
- Helping the individual to grow in all cases.
- Working to improve social conditions and benefit from scientific progress.
- The development of logical thinking.

There are lessons to be learned from the educational system of the U.S. according to the following:
- The importance of early education.
- Participation in university education.
- The quality of their specialties.
- Learning the best for larger and less expensive.
- The development of knowledge and information.
- Increasing the role that played by the school.
- Providing excellent teachers.
- Literacy as a central focus of educational reform.
- Adopting the principle of a policy of accelerated education.
- Developing model programs and standards of excellence.
- Improving the quality and feasibility of goals and curricula.
- Using all available resources including technology, non- formal education and public education.
- The establishment of boards of education to accelerate change and provide opportunities for all.

- Reviewing of the curriculum.

Features of the education system in the United State of America

When talking about the features of the educational system in general it requires a clarification of the following important components:
- Management structure and supervision of the education system.
- The structure of academic levels.
- Funding.
- The overall objectives.
- The preparation of teachers.
- Special education.
- Others.

And these components can be adopted to study the U.S. education system, where the constitute the foundations of the education system after independence, based on the transfer the responsibility for establishing the education system to each state. The educational policy aimed at establishing a form of decentralization in the establishment of educational institutions according to the following:
- Schools are not subjected to a certain religion.
- Compulsory education in the first stage, free of charge.

Moreover, there is the local council of education, who runs the education locally in each village or city in the United State. The formation of the Council the decision of the Council of the state of education and the following duties.
- The establishment and maintenance of schools.
- The appointment of teacher.
- Defining the requirements of graduate from high school.
- Identification the models of school buildings.
- Formulation of educational policy and the preparation of education budget.
- Organization of accepting students.

U.S. experience in educational planning

The local and international variables lead to pressures on education in many countries of the world and led to the emergence of reports on education in many countries of the world refers to the size of the risk caused by poor condition of education and inability to innovate and change. Perhaps the U.S. report a "nation in change of " "A Nation at Risk" in 1983 is one of the most important reports which has revealed failure of American education and called for the development and excellence, as well as a document (the U.S. in 2000 – a strategy for education), a document of Bush in 1991. As well as the document of U.S. President Bill Clinton in February 1997, which focused on the ten guidelines for Education in America for the twenty- century.

Education Funding in U.S.

The financing of education in the United States differs from one state to another and are sources of education and different funding sources, including federal, state and local levels and patterns of taxes.

Reformist experiences in the U.S.
Educational system

Several experiments to reform the U.S. education system, including the following:

- The experience of the practice of racial minorities to education.
- Reform of the nation at risk in 1956 as a result of the Soviet Union launched the first satellite.
- Education reform in the eighties.
- Reform of education in the nineties (attention to the teacher and to increase its capacity and interest in the curriculum).
- Reform of 2000 (the National Project for Development).

Futurism of education in Iraq

Experiencing the education of institutions in Iraq, major problems include teaching methods, the platform for researcher, management practices, poor use of resources and rehabilitation of the teaching staff. But the biggest problem in relation to all these are exacerbated by increases in the education of the every subject beyond the scope of the goals of rehabilitation. Curricula, programs and projects are subjected to narrow political calculations, the purpose of maintaining power and monopoly of power. It is not putting any restrictions on the circulation of knowledge only, but determine the movement of researchers within and outside the country, confiscated opinion and use educational institutions as centers of propaganda party or ideological advocacy.

In view of this low status of educational institutions, an urgent need to draw the attention of officials and alert the public opinion to the dangers of going too far in educational institutions to the interests of class and situational, deprive them of independence and the need to develop valid scientific environment to configure the format of a national scientific and independent product together. Accordingly, and based on local and regional initiatives and international organizations that have worked on the promotion of academic freedom and autonomy of institutions of education, the declaration of the UNESCO conference in Beirut in 1998 and based on what was approved by the agreements and conventions on human rights, especially the Universal Declaration of Human Rights and Civil and Political Rights and the UNESCO Convention against discrimination in education. It is suggested in the future the need to remove political tutelage of the educational community of the Iraqi and the commitment authorities to respect the independence of educational institutions in their contents of teachers, educators, students and administrators and to avoid external pressure and political interventions, which have hovered academic freedom.

The problem of global educational reform is as old as this type of activity but at the present time the problem has become self- determination in the world, especially in the third millennium, which opposes the social and economic forces and remain in the education sector in this context. But also contribute in finding a balance between the two, where it needs without a doubt to face non- traditional and a radical solution.

Education increases human knowledge, skill, consolidates the values, behavior, contribute to education in the material development, value of man and society and identify the level of civilization and therefore it is a tool of change and moral advantages of civilization. The present stage, which represents internal and external challenges, is based on an objective and realistic awareness of the problems of education, according to data accurately diagnosed as evidence of action. The education is a key component of modern states, so each state according to its capabilities and levels of its growth on the development of education due to the fact that this education expresses leadership, including:

That education reform in Iraq must be reconciled with the social transformation that occurs in Iraq, and that educational institutions should play a qualitative role in the acquisition, resettlement, the production and dissemination of knowledge and contribute to the development process in all its economic, social and cultural rights. In addition, the Iraqi educational institutions need to face the challenges of the new beyond the traditional tasks of teaching, research and service and the exercise of their duties to develop themselves to serve the social entity to achieve social goals. The fact that the future of Iraq in the near term and long- term depends on education because it is the way to the preparation of specialized manpower and the generation of thought.

The educational system in Iraq in seventies and before, achieved high rates of growth by the adoption of modern educational standards at all levels compared to the educational systems of neighboring countries. However, the deterioration of the educational system in the last two decades of the twentieth century and the stop of its growth, the quality of education declined as a result of injury to the negligence, lack of resources and the weakness of the budget and the curricula faced of underdevelopment in some respects due to adoption the political purposes.

Due to the fact that education is a national issue deal with the changes and shifts in the future, the need is now necessary for the advanced educational system as well as educational philosophy and that lead to a comprehensive change with high quality, to rise distinct levels that include the ability to do research and innovation and contribute in building education and provides a clear opportunity for all, based on clear parameters, such as:

Toward the future of advanced education in Iraq

The future is problematic and fraught with difficulties and may be related to underestimation or double precision and the lack of evidence. It is also the option depends on the scientific method reliably measured with the indicators in study. Thus that such indicator will be transferred from the planned level of metaphysical speculation to the level of scientific foresight, as well as it is past and the future for both. We need to take them into account in the formulation within the future of education in Iraq.

The educational sector is linked to all the current changes around it, requires in most cases, the prudent plan that takes their input. The plan had only two choices: either be seen as a holistic view of precision, taking into account all the actors associated with the educational process and its outputs, or slipping into the overlooked to be the wrong decisions.

Noting that the future of education is a holistic long- term effort, it is imperative to examine all the effects of past and present on the process of education, without dumping in the local and away from the world. The responsibility of this approach lies not with the education, but also on other sectors in the least. The future of education in Iraq must be linked to the guidelines associated with the input of education, manpower of the output of education, the financial aspects, the philosophy of the country, and the value system in society, drawing on all of the common principles linked to global developments and the global scientific traditions. Iraq has suffered much, painful in most aspects and consequences and are still affecting all joints of the community, suffering the education is currently in many and complex, but all of these do not reduce the motivation for education.

Visions of the future of education in Iraq

The educations believe that many of the future developments and visions of education in Iraq could be tools and products so for a successful educational strategy it should included the followings:

Once the necessary things in Iraq are moved quickly in rebuilding the educational system and its development and then re-renewing, are due to the following:

Education in Iraq faces many of the dilemmas of the future, including the continued increase in the number of students, declining levels of students in schools and inadequate sources of funding necessary to meet the education requirements due to the increasing developments and scientific discoveries and technological.

The result of these persistent problems and lack of access to critical solutions to the fears of the future of its employees in matters of education in Iraq was held during the previous years, several conferences and seminars to discuss the future of education in general. Some asserted that the time has come to reconsider the function and message of the education institutions locally and internationally and their relationship has caused a global revolution in information technology as a serious turnaround in the methods and routes and access to knowledge and was a signal one way or another the traditional role played by educational institutions in this area.

Important challenges facing Iraq in the field of education are not only to rely on research, future studies and scientific evaluation of the ongoing activities and programs at various levels. Future studies and its interest being in the developed world is of great importance on shaping the future image of Iraq, and these can be important indicators when formulating plans and strategies of development of all kinds and levels.

At a time when the last quarter of the last century, scientific developments, including all levels and fields in all parts of the world, Iraq was engaged in successive wars led to the blockade imposed by the United Nations, which has spread to more than thirteen years on the aspirations of men of science in their quest to keep pace with the scientific revolution and modern information. The legacy of this size over quarter of a century should be left on the march, in which the scientific activities in Iraq and gain from the education sector, where many abandoned scientific minds in the conditions and the weakness of low- pension means of education. As well as the poor relations with scientific institutions abroad and lack of development of legislation and laws that necessitated by the nature of the stage which led to a failure of education in Iraq very much.

This situation does not mean the frustration and despair; no one can take us to a justification to stop. Despite the difficult circumstances that followed (2003) by an unjust campaign against the Iraqi scientific minds, the threat of death, which created a new situation of panic and concern in the scientific world prompted, pushed a significant number of Iraqi scientists to leave Iraq. Since (2004) we started legislation a number of laws that would halt a sharp decline that followed it in education. An ambitious plan to lift the education of its reality, was established new tracks for the reconstruction of educational institutions with the latest technical means and modified Education Act No. for the year 1998, as amended, as well as building bridges of relationship with many of global education.

To initiate re- structuring the ministry and their scientific institutions, it required effort and perseverance that did not stop at end. Also overseeing more than 20 public education department deployed in Baghdad and the provinces in such circumstances known security. Also it required the availability of material and qualified leaders to did its part and start a campaign to fight the flab and disguised unemployment and control illegal cases. Whether at the level of teachers, teachers or students and initiate self-assessment this is what we have from the ministry and to the end of their last detailed in the school and create an atmosphere for learning environments, stems from the absolute faith that the achievement of justice and building on the principles of National Rights. The right and distribution of educational opportunities on everyone without the prior requirements, as well as the adoption of the principles of impartiality and merit of scientific improve education and access to the ranks of developed countries. The types of education in Iraq at the present time would, in principle, the cluster size of knowledge circulation and open social members with a relatively high awareness of their ambitions and their goals for the future.

In short, patterns of education developed will produce tens of thousands of people qualified in theory, but at the same time the Iraqi economy does not have the current environment and a system of current production capacity to absorb such massive numbers of learners. The new arrivals to education did not have the same time a future strategy because of the lack of form comprehensive development, and here begin the contradiction and social problems arise between people ensured their physical and time to reproduce themselves and between the developed and underdeveloped economy effected financially through the costs of registration fees and the price of the curriculum which is unable to absorb the economic borders.

Renewal of education and missions and its role in development

The education may differ as an activity and a national organizer between country and another. The tasks assigned and the desired goals may vary, but with all of these differences of the communities share the same seen as a port for the future and an eye on the givens and more capable than anyone else- from social activities- to meet those requirements.

Based on this critical role the direction of society had to be on maintenance and development, support bases and improvement of its mechanisms. These efforts must be based on a philosophy of society, non-adherent, closest, hampered by the newer and follow- up developments, preached to all the joints of the progress of social, political, economic and proposed for the renovation of ports and approaches and mechanisms are governed by multiple elements including:

he (term) renewal have been found in the field of education, became so well known from his share of the messages. But some are still confused between the words innovation and change or invention or update "Modernization" or creativity "Creativity", or reform "Reform". To clarify the privacy of any of these concepts we can suggest the following:

Futures of the educational system

In the absence of modern technology, the lifestyle and mechanisms of work in the social environment not exposed to significant change and not a comprehensive development for the production of knowledge and open the way to get to them and management. On the other hand, It is known that currents of this development was accompanied by democracy, political economic globalization, trade agreements, cultural and retail. Those had a profound impact on the contents of education in general. The expected increased need for coordination with the world to adopt global mechanisms such as: exchange of students and teachers, study abroad, study foreign languages, international studies, studies and comparison and others. It is still as long as the trend towards engaging with the word moving, it is expected an increased need for reforms in the education curricula in the field of universal values, such as human rights, democracy and the environment, and in global skills, language, computer software, such as, the independence of educational institutions.

In order to achieve effective role in the management of education, the Ministry of Education cemented relations professional, continued interaction, coordination with the community as a whole and the state and the requirements of legislation. The identification of strategies will be the principles of academic freedom and autonomy of educational institutions, foundations of those relationships required for any institution educational considers itself a social environment to investigate the performance of free and creative aid critical in the wider society.

Other proposals for the renewal of education

Priorities in the educational renewal

The difficult problem encountered by the Ministry of Education after 2003 as well as during her adoption of statements agreed upon and therefore call re- evaluation and coordination with international agencies (UNICEF, UNESCO, World Bank). A symposium was held in Amman in November 2003, jointly by these agencies as well as the World Bank. The World Health Organization and the World Food Program and the UN program for Iraq in addition to the U.S. Agency for International Development.

Moreover, the draft has been prepared and a survey in coordination with UNICEF, UNESCO and the Rise project was introduced in January 2004. All these seminars and posts provided the results of a survey of data adopted in educational information management system which was introduced later.

Then were re- organized the content for each administrative department according to the general need and objectives that has been mentioned by Dr. Alwan in his book (about a common vision of raising Iraq). These concepts may be needed in the whole state of education in Iraq, as stated by the first quarter of this document the features of the current status of education in Iraq and the roots and the history of the fords, this vital sector of society that needs to be in this period to processors complete and accurate.

The goal of the educational innovation are that the assistance of the educational institution to create positive change and to facilitate a way to move what is in fact the current traditional educational help and to meet the requirements of self- renewal in the educational system in the culture of the national community. And can represent the characteristics of educational renewal.

There are many of the mechanisms adopted to bring renewal or development, and is accomplished in the literature administrative planning. of the concepts in this regard, and the most important of these mechanisms with training and often unreliable and in an exaggerated manner to address the problems of most of the administrative organization.

In fact, that the training may contribute actively to change the behavior of workers positively, but it does not guarantee the overall change in the administrative organization or for solutions to all problems. It is clear that many of the problems experienced by the organization that had nothing to do with the human element originally, and therefore can not be trained to solve these problems.

In the case of education in Iraq and a reference to the reported concepts training is required process sharply to contribute to the following:

It could be distinguished between several types of educational renewal, renovation on the mismatch of intellectual and conceptual level, including what is on the level of material and technology, including what is on the level of the educational system, including also what is on the level of behavioral practices, or the level of human relations. These levels are also relevant to the reported elements, and can be renewal and a variety of techniques to deal with reality, including:

- Replacement of some positive elements, including the educational system of educational administration and the system of the results of examinations, or the acceptance the system for advanced students of education, the accomplishment of a new discipline, or technological equipment for laboratories as alternatives to the old equipment.
- Acceleration of scientific and technological events represents some positive changes in organizational structures and educational structures to catch up with recent developments. In the existing global fields, multi-jurisdiction includes the introduction of mechanisms to follow up very effectively to ensure or facilitate the output of scientific research or education. It is clear that this is consistent with the importance of time and

the need for administration actively in all planning and implementation activities as the investment of active time investment which is a strong evidence of the efficiency of the system and safety orientations and procedures.

- The new addition to the mechanics by the addition of some new educational elements without modifying or change in the existing educational infrastructure (research projects, studies, annular, the use of the internet). This includes enriching the educational system (administrative, leadership, research, teaching and training) and to address deficiencies in the design and mechanisms of action.

- Development a new structure is with rearrangement the work environment, or changes in the programs of study.

- The development trends and tracks and is building functional educational institution or in their functions in research, education and community service and environmental stewardship.

Iraqi educational philosophy

Sources of Iraqi educational philosophy
- Islamic faiths
- Iraqi nationalism
- National heritage
- Contemporary scientific trends
- Economic trends and community

Iraqi educational philosophy in this area based on sources can be identified as follows:
- Islamic faith and other religions.
- Iraqi national figure and development of Iraqi personality.
- Pan- Arabism nationalism and heritage.
- Political and social conditions.
- Social and economic trends in the world.
- Iraqi legislation (the constitutions and laws of educational and other).
- Contemporary scientific trends and their impact on the educational goals.
- Contemporary cultural trends.

Educational Iraqi philosophy

- The features of educational philosophy in Iraq
- Iraqi educational goals
- Curriculum and educational philosophy in Iraq
- Educational policy in Iraq, according to the proposed philosophy of education
- A new formulation of education policy
- Educational objective
- Overall objectives for the preparation of teachers

Educational policy in Iraq, according to the proposed educational philosophy

Educational policy is a set of principles, rules and criteria that determine the educational process, the key trends that define the interface movement in society towards the main goals during a specific time period and include educational policy themes, including:

There are enormous changes in the various fields in most communities due to higher level of education of the individual and society as a result of the evolution of scientific discoveries. In the field of information, electronic communications and satellites that has led to society's transition from an industrial economy to the era of the digital economy and information society knowledge and power that have become influential in the development and growth of any society that was accompanied by changes and shifts in political, economic and social challenges at the international level of the collapse of the socialist camp such as the rule of the capitalist system, reverse the role of the state, the expansion of the role of the private sector and the emergence of the so- called phenomenon of globalization.

The rapid transformations that we have mentioned its tool of education, education for change, the development of society, the nation in the preparation of generations and prepare them for future leadership and development of life. So it was necessary to a good selection of different education programs, in order to achieve the educational goals based on the doctrine of homeland, culture and needs and political trends and the idea to achieve tracking of several ways, including:

Accordingly, the computer can be used in the evaluation of education policies and educational status of the foundations of scientific- technical view of contemporary culture as to affect the education system both in terms of the contents. Of education and its relevance to the needs of the rapidly changing labor market and in response to the change made in the contents of the technological competencies required for the occupations and ways of performance or in terms of education tools. It can also assess some of the projects, including educational technology project (The Economics of Open Learning) and diagnosis of the structural weaknesses in the education system and thus put an advanced educational policy that has taken into account all the requirements of present and future and also is the most important foundations of the new Iraqi educational philosophy.

The educational policy represent of any society and, as mentioned above, determination the general shape of the structure of the educational process that allows it immunity. The principles that are going under as well as the general form of assignments and types of education and academic levels, which is governed by the students who had previously been achieving its goals and set plans, curricula, and trends adopted legislative including laws and regulations and instructions under which they operate and the general principles which are not in light of the educational process and educational policy also represents the general framework that guides the work of technical and managerial education for the device and finally it represents the general educational goals emanating from the educational philosophy that had been founded by our sources and that represent:

- Islamic faith.
- Nationality of Iraqi.
- Multi- intellectual values of Arabism and Arab nationalism.
- Contemporary trends in educational thought in the world.

Of the most important themes of human development in central Iraq and the relationship of education is the following:

- Achieve a balance between population growth and economic growth through the achievement of population growth rate.
- Providing the health care and reduce mortality and morbidity through different:

- Providing basic health care to everyone in Iraq through a system that is cost- effective, efficient and promote disease prevention and public safety.

- To provide preventive health services and emergency service.

- Reduce mortality and morbidity of various equivalent levels in different countries.

• Dissemination and promotion the care and knowledge and the development of education through the following:

- Establishment a climate that disseminate and promote the knowledge and care and illiteracy.

- Given priority to the development and dissemination of basic education, making it available to through a system that is cost-effective and efficient in order to achieve equal opportunities for citizens.

• Building an educational system based on the key disciplines needed by the national economy and provide the necessary facilities to conduct applied research in all areas of social and economic development.

• Provide a system of primary, secondary education, technical education and vocational training able to prepare labor can adapt to the needs of the labor market disciplines and skills and to achieve the level of income commensurate with their performance and productivity.

• Creation employment opportunities in both the public and private sectors to the Iraqis, training and rehabilitation, including commensurate with the needs of the market through the following:

- Creation of employment opportunities for Iraqis who want to work to avoid identified unemployment among them, with the need for training and rehabilitation, commensurate with the needs of the labor market.

- Raise the efficiency of the labor market, by bringing the benefits between the public and private sectors.

A look at the reality of education in Iraq

Improving Education in Iraq, has become indispensable at the present time, which is a new community called the Knowledge Society a society requires the capacity of mental speed, potential and special skills, face challenges that require qualitative changes in education systems.

Since education does not meet the aspirations of Iraq, with more problems day after day, the main reason is the human, and ultimately the output of education is therefore, requires the development of a strategic vision to improve future performance in the Iraqi educational institutions, depend on squarely face the challenges and problems of the education and development solutions to treat and entry points for continuous improvement. The accumulation of time of the events gave education personally, identify features in certain periods. The ultimate desired goal of student researchers from each corner of the world, passed in the age of the dark almost erased the history of science.

It is natural that restoring the country, included the midst of humanity's greatest civilizations, its health and scientific institutions stand in the position of new production of knowledge. Iraq has suffered decades of past wars and oppression, siege, heavy losses in lives and material damage. There was damage and loss that was confined to the individual, but transcended it to the community at large and even the ground won its share of that. But the biggest impact lies in the loss of opportunities for progress and loss.

The survival of Iraq in place, relying on savings from the experience at the time of the world that was in steady progress is a particular loss. Although the external form of education in Iraq over the past two decades was shown as having a proper basis, but the careful was able to see the elements of erosion and weakness of its physical and human.

Teaching have not had (in the circumstances of the siege of internal and externals) informed us of science and its innovation in the world, which produced generations of graduates with limited sight in their fields. Approaches of scientific leadership and personnel, was suffering from the constraints that made them the decisions of executive organs without allowing space for creativity, innovation and curriculum remained inherited from past eras that it became history, laboratory equipment. The new where almost annihilation, saturated conditions, some political analysts responded to the siege, others were reportedly causes to the political leadership and to its negative effects on education, agreed.

The country has witnessed an almost total breakdown in the educational system, which have never been seen before and thus, the current status of a painful education, promises continued deterioration – if effective strategic measures continue to leave the country asylum from death threats or otherwise, and devastating the infrastructure of considerable destruction.

Technology and Higher Education

New challenges for higher education are witnessed by the technological revolution and the production of knowledge as well as the economic challenges and globalization. Technological revolution helped to create a new reality in the fields of science, knowledge, information and communications, resulted in forcing many educational institutions to reconsider their curricula and economic content. Then the impact on the future or the philosophy and objectives, various educational institutions as well as management and control have been noticed.

The major global challenges have merged, clearly defined distinguished and experienced on higher education institutions. Policies, led to the introduction of significant shifts in scientific developments, social, economic.

In effect, countries are naturally divided into two categories: those active in higher education or advanced to a precedent, and those that are passive.

Education and higher education instructions produce economic public good, where academics enjoy a level of economic and working conditions of eminent there is a difficulty in governing the university by the market, but the reality of university education in developing countries, including Iraq's education structures in its content and methods, is traditional but suffers from problems of administration, organization and funding. Therefore, the first step of confrontation to operate universities is to understand the times and challenges. However, the positive effects of globalization on higher education have emerged by increasing numbers of universities linked to the internet which offers its services globally, including Jones International University, which serves students in more than 38 countries, as well as the emergence of cross-border Alliances University. More than 200 grouping of universities worldwide such as alliance Peking University and Seoul National University and the University of Tokyo and there are more than 20 university on the internet.

Globalization is vulnerable to higher education making reform of this education and the need to make it more important than ever, and that any neglect of this sector threatens the entire development process. The pressure of globalization makes it necessary to allocate adequate resources for higher education section therefore there should be reform of this sector at the enterprise level and the system as a whole.

The Higher Education is helping developing countries. It benefited from globalization through technological developments, mostly in developed countries and higher education could be a fundamental tool because it helps developing countries to reap the benefits of globalization, the latter can help countries in attracting foreign investment, and help any country products to benefit from higher education. Finally linking the globalization to higher education provides opportunities for improving living standards, India for example, benefited from globalization by building the computer industry.

The globalization of the economy includes the agenda of higher education business for the International Trade Organization "WTO". This is not for the contribution of higher education in development, but as services or commodity trade. Higher Education has become a market that generates substantial funds, as well as the globalization has affected extensively on universities resulted from the policies of globalization, low support of the public sector in a free-market economics, led to commercially-run universities. Several European universities, which depend on market cooperative economics, did not follow the way Anglo-American universities in making managed trade.

When human being was thinking of his future, his primitive tools that were addressed neither argued his aspiration nor prevented him from seeking the future. These tools were developed and became available and necessary to study and predict futurism. Later, we noticed that the scientific progress - in Chemistry, Biology, Mathematics, Computers, Medical Sciences and Engineering Science - was growing every year.

Huge developments have taken place in Chemistry, and Life science flourished new branches and witnessed tremendous development in the study of Molecular and Atomic structure, such as: the use of Lasers and low & high energy X-ray. The Chemistry of life has merged, including molecular Biology and chemical Biology that have a great future, also got tremendous developments in enzymes either in the life science and includes dozens of science specialists subsidiary, has used genetic engineering applications in areas of medical, agricultural, industrial and vaccine production and hormone treatment of incurable diseases.

The event in 1996 was exciting, the birth of Dolly in a somatic cell into a specialized egg-enriched after removing core and planting it in the womb; the most important point of this event is the return of specialized cell and embryonic stable situation after losing this status. The other development in science is the production of sufficient quantities of food in the world. Many thinkers expect that the world will see a lot of problems, related to scarcity of resources and energy, such as increased pollution and population explosion. Most studies of future ending 2025 required further means .

Future studies are focusing on several scientific areas such as renewable energy, genetic engineering, biotechnology, electronic industries, the manufacture of computers and communications, telecommunications space and material science. Researchers expect that the use of satellites for the transfer of solar energy to micro-wave stations can be broadcasted to the ground as the receipt and then be transformed again into energy that can be used, then in the field of genetic engineering, many recalled of the perceptions of future scientists in the following futurism.

Knowledge is the product of mental process of education and thinking. It is the basis of force and earnings constitute the most important components including the work and activity, particularly in the relation to economy, culture and education, which is also an economic asset. Furthermore, it includes the human being and his care and preparation of the main sources of knowledge.

Others believe that knowledge is the awareness and understanding of facts or acquire information through experience and means to acquire the unknown and self-development and is directly related to information, education and communication. The knowledge is also the most important component included in any work or activity, especially with regard to economy, society and culture.

The challenges of knowledge society emphasize the importance of university education, and this requires improvements in university education systems in order to transform them into the production process of knowledge.

Higher education is the foundation factor in the relationship with the knowledge society, which facilitates all exchange processes. The advanced stage of acquiring knowledge is the higher education which is the wider entrance to the knowledge society.

Knowledge society, which is called the community of the twenty-first century (third millennium), has the potency for production of knowledge and converting it into profitable commodities which led to the strength of this community.

These infrastructures denote the facilities, services and logistics needed by the community such as, means of transportation and means of communications.

It is worth mentioning that the knowledge society is the basis of the information society, according to the report that was issued by the United Nations educational scientific and cultural organization "UNESCO" in 2005 under the title "From the Information Society to a Society of Knowledge".

In Iraq, the knowledge society is defying Iraqi universities for being easy to be challenged, since knowledge society is advanced, and Iraqi universities are backward. Entering the university into knowledge society requires providing sophisticated infrastructure of communication technology, favorable climates of stability and contemporary education system with new techniques that emphasize the supreme actual operations.

The production of knowledge is an advanced stage that acquires knowledge and can be measured in this production through scientific publications, patents and innovations. As noted, a significant increase in the movement of scientific publication in the Arab countries, where the number of publications of the Arab scientists elevated annually from 465 bulletin per year during 1976 to about 7000 paper in the year 1995 in different scientific fields such as medicine, agriculture, basic sciences, according to the human developments report for the year 2003.

It was noted that a wide gap between Iraq and the nations of the world in the production. In order to reduce this gap substantial changes has to be carried out in the education system especially the universities, where there is a need, to build up, a modern communications network, the rules of knowledge, new information, participation in the information systems and disseminating a culture of team work education, the fingertips of the Ministry of Higher Education strides for the insurance of scientific journals in various specialized scientific disciplines and humanity.

The specialized scientific publishing gives special character followed by systematic nature and scientific method in writing, and is exposed in details to the process of scientific research and technical complexities involved and many difficulties and problems. The important point is supposed to disclose an interest in the reality of specialized scientific publishing as an essential step to improve this situation and its development.

The diagnosis of the problems of scientific research is the basis for the introduction and description of their treatment, ensuring the interest of achieving scientific research, which constitutes an integral part of publishing scientific research.

In the light of existing problems, for example, there is above all an urgent need to develop standards of sophistication and effective legislation to regulate the process of scientific publishing .

The comparison of the rates of publications of domestic and external level, in different continents, will find out that researchers of East Asia and Latin America are publishing more locally, whereas in Africa the publications high abroad. In developed countries the publication abroad does not exceed 20% for France, 25% for Japanese, and 12% of the overall European researchers.

All periodicals published in languages other than English are not worthy deserving the attention. The statistical reports that included researches from English speaking countries show that very low percentage published in other languages, while it is estimated that 17% of researchers published in English for French-speaking and 36% of Spanish or Portuguese-speaking.

Knowledge economy represents a new type of economy, different from the old economy, which was based on land, labour and capital as factors of production while the new economy adopts the so-called knowledge economy based on factors, including technical knowledge, creativity, intelligence and information.

The main forces that oppose to the knowledge economy are working to change the rules of trade and national capacity in the knowledge economy.

Knowledge economy plays an important role in the knowledge society, globalization that is dependent on the economy; it is also contributed to the result of knowledge in technology developments, according to the special requirements:

Comparison between countries of the world in research and their potential revolution, each rising in industrialized nations, falls in consumer countries for industry, which confirms the link between arbitrator and higher education.

Within the prospects of the knowledge economy, the government may support "multiple channels of higher education, including research and development" and in the area of health and medicine, in particular the increase in life expectancy during the last century 1900-2000 about 30 years, consequently resulted in an increase in the imports of society about 2 to 4 trillion dollars.

In order to play the Iraqi university prominent role in the knowledge economy, learning process is necessary as economic activity contributes essential role in the production of knowledge which is necessary in a series of knowledge society, and that the Iraqi university should resolve many of the problems.

The new pattern in higher education is to ensure the quality and its management, as well as that there is quality assessment and quality evaluation and quality assurance. The term assessment will receive many meanings and connotations, while the term "performance standard" refers to the level of achievements. There is confusion between standards and criteria, while using quality assessment and quality review as synonymous to evaluation.

The terms accreditation and quality assurance in higher education differ from one country to another. The standards are used in the USA as the same meaning of criteria, but in Europe synonymous of quality assurance; whereas the quality assurance is a part of the quality management in higher education.

Various concepts of quality in higher education are used to achieve accuracy and perfection through quality management. Quality is a unique kind of performance more applicable to higher education, as well as the quality of students, that is the ability to change constantly. Another concept of quality is the ability to report the value money and this concept has become common place, especially when something fits the quality product. There are other concepts of quality, including the concession (Excellence). Quality is suitable for the purpose of fitness as well as the traditional concept of quality of higher education, has been associated with the inspection and rejection. The transformation of this traditional concept of quality in higher education to the concept of quality assurance of higher education is based primarily on the need to test the typical rates of performance and build quality management systems for higher education. With the difficulties of application it appeared extremely important for the application of total quality management in higher education and requiring participation to ensure the survival and continuity of higher education institutions.

Quality education means an estimated total characteristics and advantages of the product to meet the educational requirements of students, labor markets, society and all internal and external benefit. The achievements of quality of higher education require directions of all of human resources, systems, methods processes and infrastructure to create favorable conditions for innovation and creativity.

Austin designs two criteria for defining quality, especially in higher education, <u>the first criteria</u>: the view that the concept of quality in higher education, must focus on the fame and reputation of the institution or its sources, <u>The second criteria</u>: is believed that the definition of quality education must be enhanced and strengthened through the application of philosophy of improving quality.

The traditional concept of quality of university education is associated with the screening operations, focus on the test, then the great importance for the application of total quality management in the higher education to ensure the survival and continuity of higher education institutions.

Quality of higher education includes indicators for measuring the level of achievement in teaching and measuring of adequacy of the needs of labour markets, the number of students to each teacher, spending per students and repetition rate. It is the concept of multidimensional and dynamic levels.

Quality is important to improve the outputs of the educational process and develop the participation of the country and society, where the country has the responsibility to adopt quality standards that must show their availability for the president of the university, dean and head of the department, may be unable to achieve on its own in certain situations.

The quality of higher education requires the direction of all human resources, policies, systems, methods, processes and infrastructure in order to create favorable conditions for innovation of educational product.

The culture of quality and its programs lead to involvement of everyone, management scientific integrity, student and faculty member, to become a part of these programs and therefore the quality is the driving force required to push the system of university education.

The phenomenon of Arab brain drain started from the nineteenth century, especially from Syria, Lebanon and Algeria. At the beginning of the twentieth century, this migration was increased especially during World Wars I and II. Migration of doctors, engineers and scientists to Western Europe and the United States until the year 1976 to about 24000 doctors, 17000 Engineer, 75000 researcher. In the last fifty years who emigrated from Arab countries between approximately 25 to 50% of the Arab workers to the United States of America, Canada and Britain attracts more than 75% of Arab immigrants. Moreover, 50% of doctors and 23% of engineers and 15% of researchers from the pool of talented of the Arab graduated last fifty years are now migrating to Europe, America, Canada and 54% of Arab students who study abroad and not returns to their countries. The Arab doctors working in Britain about 34% of doctors working there.

The losses of Arab states came as a result of Arab brain drain to about 200 billion dollars annually according to the report of the Arab Labor Organization in 2006. The Egyptians brain drain is a big problem after the figures announced by the large statistics which indicated that the number of Egyptian migrants reached 824 thousand of qualified immigrants to Europe and the United States, Canada, Australia. The scientists represent only more than 10 thousand immigrants and the rest between the fields of medicine, engineering and basic science, agricultural and humanitarian. The report released by the Arab League in 2001 warned that the Arab countries lost two hundred billion dollars due to the migration of scientists and the report noted that the Western countries are the biggest beneficiary from embracing more than 45 thousand Arab degree holders and higher qualifications.

At a time when one of the studies indicated the United Nations Program for Development to that between 1998- 2001 witnessed more than 15 thousand Arab doctor. According to statistics from the UN organization that almost 50 percent of doctors and 33 percent of engineers and 15 percent of scientists from the graduate pool of Arab States talent attract Canada and Britain, 70 percent of the migrants.

The Arab countries have measures to reduce immigration and in particular Iraq and Egypt moves, through legislation and measures to reduce the brain drain, including the issuance of Law No. 189 of the return of competencies for the year 1970, but that the calamities that have passed in Iraq during the eighties and nineties and after 2003 and thousands of Iraqis left the country with skills and qualifications. The Arab Parliamentary Union at its tenth conference in Khartoum on February 11, 2002 a number of reasons for the emergence of brain dram and capacity to absorb people of talents.

But how to deal with the file of the brain drain generally provoked a building itself, some countries has poured with extensive privileges on these qualifications wishing to return to serve their country and perhaps these met with initial success, these calls, but did not continue in its competence and on the opposite side clearly. China has a lot of scientists re- qualified as part of the world they saw as a serious political project of nation- building and can be mitigated migration.

The Arab countries, measures to curb immigrations and in particular Iraq and Egypt by legislation and measurement to reduce the brain drain, including the re-adoption of Law No. 189 of competencies of 1975. But the calamities that afflicted Iraq during eighties and nineties and after 2003 led to the leave thousands of Iraqis from the country's skilled and scientific qualifications.

However, the methods of dealing with the file of brain drain in general led to provoke the drain itself. Some countries have poured extensive privileges to these competencies, wishing to return to serve their country and possibly met with initial success of these invitations, but did not continue in its competence. To the opposite side, China clearly considered a lot of qualified scientists worldwide, in the framework of a political project, considered to build the nation and can be Immigration

Evolving economic and social development through what is being offered educated manpower, result a relationship between higher education and the economy. If we examine higher education, we find a combination of consumption and saving where the family is spending on higher education for future profit.

On the other hand, there is a set of economic challenges related to higher education that can be considered a package of economic and educational problems,

Capital representatives of the growing role of market regulations at the university may cause "the end of the university" and thus the loss of independence, leading to replace the rule of the rise of economy standards. The rise of administrators in the face of academics, the dominance of large companies and capital to work and higher education become a captive market system.

Higher education is at the forefront of the factors responsible for the economic growth experienced by developing and developed countries for the existence of technical skills, professional and specialized disciplines, where companies depend on selecting employees carrying scientific degree, and those who would perform the responsibilities on time and pass examination in unusual circumstances and adopt these data as an indicator. The test also forms a positive impact of education on economic growth.

Concerning the increasing of the quality of the work achieved, due to the positive impact of higher education, the "Asian Development Bank" in 1989, found a strong correlation between the number of years of education in the early eighties of the twentieth century and the annual rate of change in the "GNP" (General National Production) for everyone during 1965-1985 in the economy of thirteen developing Asian countries except the Philippines and Sri Lanka.

Some believe that ignorance of higher education stands behind it. This opinion is adopted by the economists who determine a simplified method for assessing the return on investment in higher education. The fundamental problem is the measurement of the return on education during the distinctions of other.

Higher education achieves the benefits of outweighed increase in income which receives an academic degree, such as participation of higher education in leadership and management culture. For example, it was noted that the export of higher education services contribute greatly to the prosperity of the United States economy in 1999. The United States gained the largest sources of higher education services in the world.

The investment in higher education represents the budget spent on scientific research. The UNESCO institute for statistics report in 2004, indicate two things, the first is Total expenditure and then Expenditure intensity. Both are increased in developed countries, and decease in developing countries.

Global Tests in Higher Education

The reaction of traditional university is in an attempt to restrict the spread of distance teaching and reduction of liberty. These universities of learning represent highly technical institutions, working in a team to develop their potentials through the production of knowledge; yet to follow the rules and regulations facilitators to work by using advanced technology in the learning process. This team is supposed to be balanced between the scientific, social and be open to the outside world.

The composition of this community is carried out through the organization of the university students requested in small groups with the supervision of teaching. These communities may appear by default and called upon the educational composition of the default and then via the internet. The advantage of this free society produces for its members where the exchange of experience among its members, a constructive dialogue and reflection in force. The exchange of experience and knowledge, and therefore the virtual community, is opened to all individuals of all societies and in light of this growing knowledge capacity of associates to improve their knowledge and skills and patterns of thinking.

Institutions of higher education in America are classified to four main types, of small colleges, institutes of technical, liberal arts colleges and universities. In turn are divided into two types, government funded by federal government and by another type of private funds.

The British universities are estimated to be at about fifty universities, Oxford and Cambridge are the oldest established universities. They were founded in the twelfth century and the method of teaching in these universities is called the Pilot Tutorial Manner. There are other universities including the famous University of Wales in Wales, and Edinburgh, Aberdeen and St.Claskwo and 22042 in Scotland.

Universities of British traditions stemming from the universities of Oxford and Cambridge have changed to some extent, the development of social changes have become many types and patterns of higher education.

The French Universities are regarded as centers of university traditions, including the University of Paris and a number of universities around supervised by the Ministry of Education.

There are different types of university education in France such as traditional French university of state-funded.

University Doctorate regulated the same university and Doctorate state grants for scientific research with the original level.

These universities grant various degrees, including the first university degree. High Diploma Studies take between 4-6 years degree courses, the equivalent of science Doctoral degree that requires three years of study. The Doctor of science, which is the highest degree, was issued in a decree in 1990 to reform universities with greater autonomy and broader authority for universities.

The challenges of Iraqi universities

Iraqi universities face serious challenges including challenges associated with internal structure of universities, philosophy, methods of science, factors governing the forest and the quality of universities. External challenges are associated with knowledge society and Iraqi universities face these challenges, internal or external so it has to reconsider the philosophy and goals of higher education and activating the role of Iraqi universities in partnership with community. The interest of the quality of graduate, working on deepening the Iraqi identity in universities developing a new vision and a message to achieve this vision for tertiary education and establishing a strong infrastructure of modern communications network and the introduction of tests.

Iraq has been living at the present time one of the darkest periods throughout history, where it faces serious challenges, linked to its ability to survive, and affect the security and stability, and even threaten its existence and entity.

The challenges facing Iraq are many and varied and overlapping, making it vulnerable to military occupation. Great States came to Iraq pretext of freedom, democracy and human rights, which claim the right to be invalid by the pillaging, looting, and vandalism, and humiliation, and if the basis of the challenges facing Iraq is the failure of different dimensions and forms. The distinctive way through which to address these challenges is to develop university education and leap from the reality of underdevelopment in all its aspects and limitations to the reality of progress in all its dimensions and prospects.

The university education carries the most important responsibilities of: preparations, development, improve performance, and transferring Iraq from an under-developed state to a state in the phase of progress. The university education, like the Iraqi society, is facing many problems that shackle its movement, and reduce its role in improving the performance of Iraqi rights.

Prospective of Higher Education in Iraq

Over the last few decades of war and sanctions, Iraq has suffered great losses in death and damages, and inflicted only on persons, communities, communications, institutions, and land, but equal and perhaps greater damage has been the outcome of the lost development opportunities: stagnated communities, uneducated children, adolescents and youth, neglected land, accelerating mortality rates, and the construction that did not take place.

The nation has also witnessed an unprecedented devastation and a near-to-complete break-down of the education system. The prevailing state of the affairs of higher education is disheartening and presents a virtual collapse; academic staff continues to leave for fear of mounting threats and insurgency, teaching being stopped or erratic and most buildings being damaged or destroyed.

The education system is currently in a stage of profound rebuilding and restructuring particularly at a time when the demand for higher education system to address the national needs and priorities comprehensively seems to be a source of much pressure. This mounting pressure is equally evident on the Iraqi institutions and centre of knowledge to produce relevant and appropriate knowledge at an unparalled rate. The diverse demands for qualified human resources of the complex national economy with evolving global links require knowledge for decision-making and for production.

Yet the challenges that Iraq faces cannot be solved with piecemeal changes; but strong forces that will affect Iraqi society in the future; economic reconstruction, population change, globalization and the rapid growth in knowledge. The situation must be met by a fundamental re-orientation of higher education activities under a unified perspective (new version). Providing this "new vision" and a "strategic plan" for specific and realistic short and medium term choices is the aim of this paper. The vision and the strategic plan have been formulated to help guide the initiatives and to provide clear priorities for an efficient use of available resources.

These include, supporting higher education process, which serve Iraq and its students and people, providing higher education for each Iraqi individual, upgrading higher education to imply levels of advance and entering to new century with fixed and clear steps. Furthermore, the ministry aims are to prepare and create scientist and specialists and to achieve the justice and equity among all individuals and society regardless their ethnic or nationality.

The strategy for the future should aim to: establish a stable and sustainable financial base, ensure effective leadership and management at all levels, maximation of the staff potential, and promotion of collaborative research.

Furthermore, to ensure the growth of a healthy and productive university research environment, the ministry of Higher Education and

Scientific Research should undertake a critical review of the universities, and should be encouraged to seek international funding for research.

During that, the policy of establishing a university in each governorate responded to both the demands of equity to both the growing demands of higher education.

Thus, there is wide range in the size of universities as well as a lack of geographic equity, in their distribution across the country. Rehabilitation of universities, colleges, according to international levels is required.

The major fields of study offered by the universities are: education, arts, law, social science, administration, economics, natural science, engineering and technology, medical sciences, veterinary medicine and agriculture.

Higher education policy and strategies

The main issues needs and priorities that emerged from the UNESCO Roundtable were the following:

- The wide spread destruction of the infrastructure of the higher education system.
- The unstable and dangerous environment for normal academic activity.
- The quality of higher education has been steadily deteriorating since the imposition of authorarition rule in Iraq.
- There is a need to equip more than 2000 scientific laboratories and for 30000 computers, libraries are in a poor condition and are in urgent need for restocking with new books and journals in both Arabic and English; journals in electronic format are required.
- The student population has been rapidly increasing due both to a high birthrate and an admission policy that allows all students who have completed secondary school to enter higher education.

To reconstruct the higher education, it is required to plan spanning (5-10 years), maximization higher education graduation, learning students opportunity according to the needs, development a team to inspire student, teachers to work in team, curricula should be based on appropriate outcomes and capabilities, development of teaching materials, class size should be designed to suit the student learning process, and the enrollment polices need to aim at recruiting enthusiastic students.

Consideration needs to be given to the relationships between higher education and industry and commerce. The need to update the information in higher education, information and learning centers should be established. A creative policy is needed for the provision of learning resources.

New Vision and academic covenant

The "New Vision of Higher Education in Iraq should be based on a judicious balance between the preservation of those features which should remain as part of the educational and cultural heritage and the changes which are essential to preserve the role society accords to higher education. The aim should be to make higher education more responsive to the general problems facing Iraqi society and the needs of economic and cultural life, and more relevant in the context of the specific problems of a geographical region and population group. The universities and institutions of higher education should be seen as:

- Institutions contributing to socio-economic change and helping to promote sustainable human development;
- Institutions contributing to the organization of modern and be more closely involved in actions aimed at reducing poverty, protecting the environment, improving health care provision and nutrition, promoting the principles of civil society and development other levels and forms of education.
- Institutions responding to the changes in the world of work and civic culture meaning thereby developing academic and professional qualifications as well as civic and personal qualities.

The above points can be summed up as follows: what is and what should be the role of higher education in present and future Iraqi society.

The new policy directives should envisage Iraqi universities and higher education institutions as developers of both responsive and proactive attitudes towards the labor market and the emergence of the new areas and forms of employment. It should pay sufficient attention to changes in major market trends so as to adapt curricula and the organization of studies to shifting circumstances and thus should ensure greater chances of employment for graduates.

The principles, on which quality and accountability in governance and management of higher education institutions are based, should be the part of the new vision. The proper degree of controlled by role institutional autonomy should be ensured for both public and accredited private higher education institutions to allow them to be relevant and perform their creative, reflective critical functions in society. Further, institutional self-governance should be given adequate and pragmatic form.

Academic freedom and institutional autonomy should imply increased responsibility in academic work, including its ethical context, and in matters of funding, self-evaluation of research and teaching and a constant concern for cast-effectiveness and efficiency.

Analyses of the present conditions of higher education in Iraq are unanimous in pointing to insufficient financial resources as one of the major constraints on its future development. The challenge of limited resources is unlikely to be overcome in the near future, so higher education institutions will have to find ways for coping with this challenge. Elimination of weaknesses in governance and management is paramount in this process. Therefore, it would be in the interest of higher education (public and private) to consider the issues of evaluation and quality, including institutional and programme accreditation, as vital for a responsive and accountable system of governance and management.

Another element that deserves special consideration in the new vision is the renewal of teaching and learning, i.e. issues of content and delivery. In order to achieve this objective, MOHESR should review and re-evaluate the place of teaching in its overall mission and provide incentives which would better reflect the current importance of this activity. This would imply a continuous analysis of the need for study programmes, training and re-training, and would require the establishment of methods for academic recognition of work experience relevant to student's academic work and to teaching qualifications. As well as local experts among the teachers, experienced people from the world of business, government, and international organizations should be involved to inject new ideas into study programmes.

The vision should clearly recognize the fact that most viable institutions of higher education, in both financial and operational terms, would be those which would succeed in incorporating mechanisms and information systems that enable them to remove mediocrity and guarantee quality of teaching, research, and service. These would be the institutions which would stand better chance in competition to obtain resources from the public and private sectors. Thus, a key to improved governance and management would be the confidence-building in the leadership and the managerial qualifications of those involved in these activities. This would imply improvements in:

1. Selection and assessment.
2. Enhancement of appropriate training.
3. Development provisions at the system, institutional and department levels.
4. Promotion of research on higher education which should be regarded as an important "knowledge-base" for policy-making.

For the renewal of higher education, the new vision should envisage the higher education system of Iraq as:

- A system for high quality training, enabling students to act efficiently and effectively in a broad range of civic and professional functions and activities, including the most diverse, up-to-date and specialized ones.
- An academic system to which access is possible primarily on the basis of intellectual merit and of the ability to participate actively in its programmes, with due attention to ensuring social equity.
- A community fully engaged in the search, creation, and dissemination of knowledge, in the advancement of science, and participating in the development of technological innovations and inventions.
- A system of learning founded on quality and knowledge, which generates particularly in the minds of its future graduate, commitment to the pursuit and emphasize of knowledge and a sense of responsibility to place training at the service of social development.
- A system that welcomes updating and enhancement of knowledge and qualifications as part of institutional practice and culture.

- An academic community in which cooperation with industry and the service sectors for the economic progress of the country is encouraged and actively supported.
- A system, in which important local, regional, national and international issues and solutions are identified, debated and addressed in a spirit of learned criticism. A system in which active participation of citizens in the debates on social, cultural and intellectual progress is encouraged.
- A system to which government and other public institutions can rely on for getting scientific and reliable information which is increasingly being required for decision-making at all levels, and which also promotes public participation in the decision-making process.
- An academic community whose members, being fully committed to the principles of academic freedom, are engaged in the promotion of human rights, democracy, social justices and tolerance and in building a culture of peace.

Relevance of higher education to national aspiration

The relevance of higher education in Iraq will be seen and considered primarily in terms of:
- Its role and place in society.
- Its functions with regard to research and the resulting services.
- Its links with the world of work in a broad sense.
- Relations with the State and public funding.
- Interactions with other levels and forms of education.

These relations will give due recognition to the importance of academic freedom, decentralization and institutional autonomy as the principles underlying all efforts to assure and enhance relevance. Relevance will not be viewed strictly in academic terms; rather the need for relevance will be gauged and valued in terms of new dimensions and greater urgency as the present economy of Iraq requires and will continue to require for the medium-term graduates able to constantly update their knowledge, learn new skills and qualities to be not only successful job seekers but also job creators in the labor market.

For the successful functioning and effective management of higher education, good relation, interaction, dialogue and debate with the State and the society as a whole (policy orientation, formulation and legislation) will be stressed at all levels of administration. The principles of academic freedom and institutional autonomy will be the basis of these relations which are vital for the preservation of any institution of higher education as a community of free inquiry, able to perform its creative, reflective and critical functions in society. While the MOHESR will assume catalytic and regulatory roles, institutional self-governance in higher education will prevail. At the same time, the entire socio-economic environment will compel higher education institutions to build and develop ties and linkages with MOHESR and other sectors of the society, and to accept that they are accountable to society in general.

Recent advances in science and technology have not permeated and percolated fully within the institutions of higher education of Iraq simply because of the country isolation from rest of the world for nearly three decades. In the absence of modem technologies, lifestyles, work patterns and changes in the social environment have not undergone noticeable transformations on the one hand and in enhancing fully the production, management, dissemination and access, and control of knowledge on the other. Along with these national trends and developments, have occurred international events such as political democratization, economic globalization, regional trade agreements, social and regional polarization, exclusion of various social groups, and cultural fragmentation have occurred which have direct bearing on the content of higher education.

In light of these global trends and post-war challenges, the fundamental question facing MOHESR is: what role must Iraqi universities and institutions of higher learning play in both short and medium term? The most logical response to the above challenge appears to be that the mission of all higher education institutions in Iraq be based on societal and individual needs, both present and future. In doing so, a clear stress will be put for each higher education institution to achieve and restore a balance between apparently conflicting forces such as:

- Technological modernity vs. cultural preservation

- Internationalization vs. the local obligations of universities
- Individual development vs. social cohesion
- Pursuit of knowledge for its own sake vs. direct service to society
- Fostering generic skills vs. providing specific knowledge
- Responding to employment system demands vs. shaping the world of work proactively
- Generalist vs. specialist training

Although the push and pull of these compelling and opposing directions are relatively difficult to resolve, the term "international higher education" will be used as an "umbrella term" for the various institutional programmes and activities that are international in nature, such as student and faculty exchange, study abroad, international development activities, foreign language studies, international studies, area studies, joint-degree programmes and comparative studies, among others.

Since the globalization trend ahs been leading to increasing social aspirations and expectations for the need of global education, the challenge of developing in the university students' global values (e.g. morals, ethics, human rights, democracy and environment), global knowledge (of ourselves and others), and global skills (e.g. language, computer expertise) will form the major content areas of higher education curriculum reform exercise.

The reality of Higher Education in Iraq

The improvement of higher education in Iraq has become an indispensable requirement for the time being. The new society in the so-called "Knowledge Society" requires great mental capabilities, potential and special skills, to face challenges that require changes in the quality of higher education systems.

Because higher education does not meet the aspirations of Iraq where more problems occurred day after day, the main reason remains the human being, which ultimately represents the output of higher education. Therefore, it requires a strategic vision for improving future performance in Iraqi universities that depends on obedience to face the challenges and problems of the Higher Education and develop solutions to address them, and input of continuous improvement. The meaning of this idea is that the accumulation schedule of events gives Higher Education personality features in certain periods.

Baghdad was the center of researchers in the world, passing through the eras of the dark history of science that is almost indelible. It is natural to regain the country prosperity annexation of the greatest civilization of human health and scientific institutions to stand again in the position of producing knowledge.

Iraq has suffered over the past decades of wars, oppression and siege which caused enormous losses in lives and material damage. Damage and losses were not confined to the individual, but surpassed them to society at large; and even land won its share of that. But, the greatest harm and loss lies in the loss of opportunities for progress. The survival of Iraq in place, relying on the expertise remained exclusively the time when the world was in steady progress, is a particular loss. Although the layout of higher education in Iraq over the past two decades had shown that it was based on sound foundations, but the elements of erosion and weakness of the physical infrastructure and human resources was clear.

Teaching staff, in conditions of the siege of internal and external, were deprived from updated information and communications in science and developments in the world, which has produced generations of graduates with limited sight in their specialties and scientific approach. Cadres and leadership were suffering from restrictions that made them the organs of executive decisions of superstructure without allowing them creativity and innovation. The curriculum has remained under an inherited form past decades until it became history. The laboratories and equipment were expired, and the new were almost indelible. Conditions permeated by some politicians analysts responded to the siege, other causes due to the political leadership, punched their negative impact on higher education.

Iraq has witnessed a collapse almost universal in the educational system; a collapse which has not been witnessed before. Thus, the present situation of higher education is painful. And there signals on continuing deterioration if the appropriate measures are not taken by effective strategy. Academics are continuing to leave the country to seek asylum abroad, away form death threats or otherwise; and the infrastructure is destroyed significantly.

Nevertheless, the educational system is currently undergoing a major phase in the rebuilding at a time when rising labor market needs to increase demand for education. These are shown clearly by the growing pressure in the scientific institutes and centers that are to produce knowledge compatible with those requirements.

The university status quo and its chronic forms were not the result of today, but relates to the accumulative situation that reflects some of the failures of universities in achieving educational and research functions at service level required. Therefore, the fundamental issue at the Ministry of Higher Education and Scientific Research no longer lies in renewal or non-renewal of the university. Renewal reality inevitably spurs better traditional solutions to the event. But, the fundamental issue is how and in the time frame needed to conduct such renewal and development.

The central issue, which clearly shows stalemate with the worsening conditions in the atmosphere of traditional university, is a prelude to create the University of the Future. This university, which is characterized by university autonomy, academic freedom and management firm capable of taking initiatives, is supposed to address the changes that take place in the structure of scientific and technical knowledge in the modern era. Such ambitions would not be far-research that elusive plans to achieve strategic planning carefully.

In the light of documents and statistical data on various aspects of university, such as admission process, number of students, teaching staff, their proportions, and the evolution of spending in the budget, development in areas of the university could limit the posts as follows:

• The area of educational function (in the transfer and dissemination of knowledge).

• Research area (production and generation of knowledge).

• Community service area (environment, development, dissemination of knowledge applications).

• Area of the university posts in the structure of scientific knowledge (and the accumulation of knowledge of mankind global landscape).

Facts about the development of the present of Higher Education in Iraq

The reality of higher education is described as follows:

• Higher Education institutions in the Kurdistan region are still administered locally; with weak coordinating bodies with the central Ministry.

• Iraqi university evening studies are still opening where most students pay nominal wages, and are committed to the same curricula .200

• The Ministry's policy is to accept all tracking graduates of secondary schools in Iraq. The vocational schools accept a limited proportion of the graduates.

• A variety of research centers which are linked to universities, differ in the performance research level.

• Various centers and offices that provide advisory services to the community are linked to the Iraqi universities.

• Financing official university institutions is still central.

• There is one specialized university in engineering in Iraq.

• The Ministry of Higher Education recognizes a number of private educational institutions according to clear and specific scientific criteria.

• The official higher education is still free of charge.

• Higher education institutions lack the effective training policies of workers despite the opening of such centers in the universities.

• Most universities lack the active systems to evaluate performance.

• Universities lack the local network (or global) for search and communications.

• Many scientific curricula are still away from the latest developments of science.

• The higher education lacks clear policies to deal with the surrounding social environment in relation with employment and coordination regarding the curriculum.

• The Ministry still perpetuates communication and direct coordination with other ministries concerned of higher education.

Futurism of Higher Education in Iraq

Problems of futurism

Higher Education in Iraq is facing many future problems, including the continued increase in student population, declining standards of students in universities, and inadequate sources of funding needed to meet the requirements of educational and scientific research.

The result is the continuity of these problems, and lack of access to critical solutions which led to the prospective fears of workers in higher education in Iraq. Several conferences and seminars were held during the past years to discuss the future of universities and higher education in general. Some of these conferences asserted that the time has come to reconsider the function and the message of higher education institutions locally and with its relationship to the world. Information technology revolution has brought a revolution of methods and access to knowledge. It is a signal to the traditional role played by universities in this area, and the institutions have shown the name of (tertiary education) which refers to doubts about the appropriateness of universities. Their composition, arrangements, regulations and rules are set to cope with the variables of the current era of accepting the new generations to rebel against all the systems, and be free from all restrictions. Accordingly, it requires from universities to take actions in the future to match with these variables. The important challenges facing Iraq in the field of higher education are not to rely on research, future studies and scientific evaluation of the ongoing activities and programs at various levels. The studies and interests in the future, as being in the developed world, are of great importance for determining the features of the future of Iraq. This could be a useful indicator when developing plans and development strategies of all kinds and levels, while it has witnessed from the last quarter of the passing century, very important scientific developments that included all levels and fields in all parts of the world. Iraq was waging consecutive wars that led to the siege imposed by the United Nations, which spread on more than thirteen years.

The heavy legacy of this size, which exceeded quarter of a century, must be left on the march of scientific gain in Iraq and was affecting the higher education and scientific research. Then, many advanced scientific minds were abandoned by virtue of substandard conditions such as: weak pension, weak means of education, and weak relations with the scientific institutions abroad. The development of legislation was so weak, and laws were required due to the nature of the phase. All those led to the failure of higher education.

This does not mean a justification for frustration and despair, nor can it be as a reason to stop at the point of launching ramp. The year (2003) witnessed a great change in the regime and the state structure. An unjust campaign was launched against the Iraqi scientific minds, and death threats created a new situation of panic and anxiety in the scientific community, which forced a significant numbe2scientists to leave Iraq. Since (2005) we started legislation a number of laws that would stop the sharp decline of the wheel of higher education. We put an ambitious plan to lift higher education from its bitter reality. It included new tracks for the reconstruction of educational institutions and equipping the latest technical means; the amendment of the University Service Law No. 142 of 1976; the rate of bridging relationship with many universities and launched the Global Program of scholarships after a break of more than 25 years.

The re-launching of the Ministry structure and scientific institutions requires effort and perseverance which can not stop at an end. Also, oversee more than 20 universities which were deployed in Baghdad and other governorates in such crucial security conditions. This requires the availability of financial resources, qualified leadership to do its role, start a campaign to combat the slack underemployment, and control illegal cases, whether at the level of teachers or students. To be on solid ground line, the Ministry has initiated a self-correct campaign starting from the ministry to the end of the latest details in the university or institute or department. Then the Ministry created an atmosphere of sound educational stems. From our belief absolute justice is built on the principles of national rights, distribution of educational opportunities for everyone without prior requirements. The adoption of the principles of impartiality and scientific maturity will raise the level of higher education and provide it access to the ranks of developed nations. The types of higher education in Iraq at the present time (governmental private, evening studies, day-time studies) will in principle, increase the size of mass circulation cognitive socially, and produce individuals with relatively high awareness of their aspirations and their future.

In short, patterns of education will produce thousands of qualified people in theory, but at the same time the Iraqi economy and its current environment and production does not have the capacity to absorb such large numbers of learners and new entrants to the labor marker. Also, it does not have at the same time futuristic strategy to absorb those graduates because of lack of the comprehensive development model. Then, there is the problem of the social contradiction between the people procured, and the underdeveloped economy that absorbs them materially through costs.

The future of higher education and university needs a number of key skills such as:

- Adaptive capacity and flexibility.
- Ability to deal with rapid changes.
- The ability to transfer ideas from one area to another.
- The ability to change Orientalism and preparing for it.

The (Deloris Commission) that have been developing (Education for the Future) program commissioned by the UNESCO raised fundamental questions about the future of higher education:

The first issue: the ability of educational systems to become a major factor in development.

The second issue: the ability of educational systems to adapt to new trends of society.

The third issue: the relationship between the education system and the state.

Fourth question: spreading the values of openness and mutual understanding with others.

Future direction of research and development

There are requirements and future directions that depend on several issues including manpower, funding and legislation, infrastructure and communication interfaces and networking. The manpower needed directions in the future such as:

- Urging staff members to attend scientific conferences.
- Hosting Arab and foreign researchers to create new research channels in our universities.
- Adoption of global standards such as students to teachers ration.
- Development the spirit of team work in scientific research.
- Documenting research work.
- Allocating University Awards to honor distinguished researchers annually.
- Urging the scientific departments to hold scientific conferences locally and globally.
- Permitting the opportunities for graduate students to spend a distinct time in other local or external universities.

In the center of finance, orientation requires:

- Increasing spending on scientific research to the rate of 1% of the GNB.
- Developing appropriate mechanisms to activate the role of fund to support scientific research.
- Cooperation with universities from various countries around the world.
- Stimulating the private sector to contribute to financing scientific research in higher education institutions.

In the area of legislation:

- Requiring future plan for scientific research on the state level in the light of the priorities of development plans.
- Establishing tight contact between state institutions, private sector, institutions of human development and social and economic development provisions.
- Submitting rules to the binding private sector, and the productive enterprises, and universities.
- Developing academic legislation on promotion and loan and a full-time teaching and others.
- Commitment to ethics of scientific research disciplines.

The infrastructure of the scientific research requires scientific future orientations to strengthen laboratory equipment processing, modern techniques developed for the establishment of central laboratories in each university, and establishment of research centers specialized in the various fields.

Past and Current Status

Modern universities in Iraq were established in the second half of the last century, beginning with the University of Baghdad in 1957 uniting several constituent colleges in the process, during the 1960s five more universities were established: the University of Technology and Al-Mustansriya University in Baghdad as well as three universities in Basrah, Mousal, and Sulaymania. During that, the policy of establishing a university in each governorate responded to both the demands of equity and the growing demands of higher education. Thus, there is wide range in the size of universities as well as lack of geographic equity in their distribution across the country. Rehabilitation of universities, colleges, according to international levels is required.

The major fields of study offered by the universities are education, arts, law, social science, administration, economics, natural science, engineering and technology, medical sciences, veterinary medicine and agriculture. In the area of education there are 24 colleges preparing teachers for secondary schools, 7 colleges for primary and kindergarten, school teachers, and 7 for physical education. The University of Technology has a specialized college for technical education, training teachers for vocational schools and technical institutes. The university education duration, to which student are admitted after the secondary stage, ranges from three to six years. Applicants are registered in colleges to pursue their studies in various specializations such as arts, science, medicine, engineering, etc… Bachelor's programs leading to a Bachelor's degree are awarded in Architecture Engineering, dental surgery, pharmacy and veterinary medicine and surgery. Programs leading to Bachelor of Medicine / Bachelor of surgery require six years to complete.

Chapter seven
Technological policies

Preface

The availability of financial resources in Iraq, especially before the nineties of the twentieth century has made great efforts in manufacturing, especially in the field of Military Industrialization. Iraq succeeded in building an independent military industrial base by enabling technology in modern manufacturing processes, however, strategies and manufacturing policies was clean from effective action to develp local capacity and providing appropriate incentives for local people to be able in modern industrial technology.

Despite this, Iraq still needs to pursue innovative methods to meet the daunting challenges of a large number of other sectors of production and services, as well as to develop scientific and technological capabilities of local traditional industries to modernize and to address a variety of social economic problems.

The scientific efficiency in Iraq itself has a capacity of innovative crucial role in facing challenges, in spite of standard conditions faced by the departure of many of them outside Iraq.

The limited scientific successes that have occurred in Iraq are the result of the efforts of some institutions of science and technology (Scientific Academy, the House of Alhikma and the institutions of higher education). On the other hand, important achievements were made in building the institutions mentioned, as well as in human resources development, but the institutions mentioned and others are still far from an enabling a distinguished role in development. The linkages and synergies between the scientific and technological institutions of governmental organization and the business world remain weak, despite the existence of some versions of contracting.

The spending on research and development in Iraq, at best, less than its counterpart in the Arab countries and more discouraging regarding the outputs of science and technology, so that scientific and technological publications in specialized areas and the number of patents granted to institution and individuals are much less than the average of the corresponding figures in Other developing countries.

As a result, the status of science and technology in Iraq needs a lot of attention. The input and output of scientific and technological point indicate the deficiencies in information networks, computers, advanced equipment, scientific research.

These include, institutions (universities, research centers universities, atomic energy, etc.), their shortcomings in Iraq are one of the reasons that led to the absence of a scientific and technological policies and the recognition of the limitations. The ineffectiveness of administrative practices and the existence of structural deficiencies are symbolic recognition of the need to develop capabilities in the next decade to operate in an environment different from recent years.

These institutions are currently working on:
- Production and dissemination of scientific research.
- Rating knowledge, covering a range of disciplines and areas of application.
- Training of some ranchers.

And, therefore it requires a complete reform of these institutions revitalization to support them, and identification of high-level priorities by increasing the competitiveness and environmental compatibility, creating an effective system of funding policies and linking them to industry and social and economic activities.

The search on sciences started with the beginning of man himself. The primitive scientific knowledge was transmitted from one generation to another through various professions. During the late nineteenth century the scientific education was dominated by religious, descriptive character and centered in the church. The science education in the early twentieth century was characterized by an increase turnout with those of social and economic modest level with scientific study. But, during the mid-twentieth century, emphasis has been used to enrich the life of scientific research, according to the needs of society and the role of science as a social power.. The learning of science should deal with many available factors such as principles and conditions, including early education and develop the ability to use scientific method to solve problems and the acquisition of the scientific education. The science education should have a role in the vital revolution, the acquisition of facts, the acquisition of functional concepts of the scientific principles and skills to solve problems and the development of scientific trends.

The development capacity and the development of personnel skills have witnessed successful experiences in building human being in the world.

Human beings play a distinct role in the technological process and the educational system due first to the human activity and second to the component of humanitarian and social structure. Technology in general inherent rights throughout the human being's life - the life of humans or living organism possess, what is referred to by: thinking, feeling, and the ability to speak, despite the existence of opinions and philosophies on the relationship between body, mind and soul. All the activities related to human life building represent a coherent body with a specific mental condition and functional.

Despite these perceptions, the human is building various scenarios in the philosophies of public and educational philosophy. The human being in ideal philosophy is exercises his ability and responsibility for his actions, whereas Plato stated the man consists of two parts, one belongs to the world of ideals (self) and the other to the world of sense (the body) and stressed on the idea of bilateral. While realists see that man is an organic entity with social tendencies. The natural philosophy suggests that the human soul is good, the present is the origin of the evolution of the future and education is necessary as long as they continue to longevity.

Al-Ghazali as a leader of the Islamic philosophy indicates the interdependence of the dimensions of human nature (body, mind, soul, heart), as well compatibility and balance.

There are several theories to explain the emergence of life in trying to answer the classic question. How did life begin? The answer is as such: the fall of some organic molecules on Earth from comets, and planting the land with life by the intelligent beings on advanced planets. The emergence of life after eons of consecutive nomination (years) with quick chemical composition following the ground composition .Then after a short time period (the escalation of water droplets to the surface followed by collected chemicals, turned quickly to life in other place, then moved to the ground). After the general review of ideas about the evolution of life, the question was raised about the secret of inductive creation of the universe and the concordance between the creation of the universe and life.

There are clear trends in the study of the secret of creation of the universe and life: firstly it is believed that there is an aim in it and secondly there is a deny of the existence of the end, but to the existence of chance. Among the topics that are interesting mystery of the creation of the universe and of life, origin of the universe, evolution and various theories that talk about evolution, which accept both science and reason, scientists and philosophers. As well as the emergence of life and theories of ancient and modern, material and ideal, west and east that deal with them. The evolution has contributed to clarify the continuing march of life regardless of the theories put forward, including Lamarck and Darwin.

The science and thought in the understanding of life

After clarification the thinking part of life, we could say the deeper main reason of the universe and the world in general is the reason due in particular, which ends by the sequence of causes and the only question that deserves its generation is due to this reason in particular, which is the first fountain of existence, is it the same material, or something else over the border?

The philosophical trend of the mentioned concepts raises a question of whether the reason is the driving force of the world, is the same material and whether the secret of life is something else, beyond the limits of material and different from them. The philosophical conflict defines the direction of the secret of life; it requires not to confuse between the scientific material and the philosophical material.

Each community the public and the private has a scientific legacy, with various pillars such as cultural, social and preserving the heritage is a national and human necessity. The scientific education can contribute to the benefit of the scientific heritage. So it is really important to stop a bit to show how to deal systematically with this legacy, thereby the process of transforming extrapolation of tradition and its benefits to the renewable power of the development of general education. There are methods and appropriate methodologies used for the purpose of benefiting from inherited science, including:

- Scientific heritage is treated with inherited methodology based away from the indiscrimination and excessive, damaging to these inherited disciplines.

- Scientific heritage depends on inherited as a single bloc in terms of time and the researcher does not fall blackouts, at the wrong focus on specific topics.

- The researcher shall comply with the spirit of scientific criticism of the inherited and not influenced by intolerance and prejudice.

- Selection process in the inherited case depends on modern scientific techniques.

Indicators and trends in technology polices in Iraq

Iraq made over the past decades considerable progress in several areas, has increased the resources allocated to technology, social services and infrastructure which has had a positive impact on the average per capita income and quality of life, followed by problems of concern (during the nineties). The per capita of the total GNP was decreased, the blockade halt efforts to diversify the economy and the adoption of the main sources of GNP on non- renewable mineral wealth.

There are some hypothesis for technological policies in the country which are limited in scope, effectiveness and ineffective strategies in the best of circumstances. So it was proposed to create new structures after some test on developed countries and developing countries.

The spending on research and development in Iraq, at best, less than its counterpart in the Arab countries and more discouraging regarding the outputs of technology, so that scientific and technological publications in specialized areas and the number of patents granted to institution and individuals are much less than the average of the corresponding in Other developing countries.

As a result, the status of technology in Iraq needs a lot of attention. The input and output of technological point indicate the deficiencies in information networks, computers, advanced equipment, scientific research.

These include, institutions (universities, research centers universities, atomic energy, etc.), their shortcomings in Iraq are one of the reasons that led to the absence of a technological policies and the recognition of the limitations. The ineffectiveness of administrative practices and the existence of structural deficiencies are symbolic recognition of the need to develop capabilities in the next decade to operate in an environment different from recent years.

And, therefore it requires a complete reform of these institutions revitalization to support them, and identification of high-level priorities by increasing the competitiveness and environmental compatibility, creating an effective system of funding policies and linking them to industry and social and economic activities.

The current levels of scientific and technological cooperation between Iraq and Arab countries are very low, as well as with foreign countries which are almost non-existent as illustrated by the lack of joint scientific research and the outcome of joint publications. This refers to the need for Iraq in its quest to strengthen their technological interdependence.

To enhance communication and cooperation between Iraq with other Arab countries is a prerequisite in better determining the technological problem and to obtain acceptable return of the available knowledge. Lounge of the basics of scientific cooperation is to avoid free turn key contains from any type of Technology, which is almost a priority involving technologists from two or more research fields, and incur the available possibilities to researchers to attend technologists meetings and make more use of international organizations.

The term technology transfer indicates that the technology is acquired through the transfer of goods or services technology and therefore is not equipped for transportation and technological deficits.

The experience of Iraq in the transfer of technology is varied and appropriate in a large proportion, completed with foreign companies which provided comprehensive and complex technological deals in the framework of international market strategy, and the country suffered from indiscrimination of transfer that took place in the absence of a sound domestic policy in various technological fields.

There are two problems facing Iraq, the first is concerning the search, transmission, absorption, development and improvement of modern technology, and second is related to technology development.

Technology policies in Iraq

Technology play a crucial role in shaping the challenges faced by individuals, organizations and nations constantly, (discoveries of genetic engineering, industrial human parties, mobile phone, biotechnology). The challenge facing mankind at the beginning of the twentieth century is how all countries could benefit from the strength of technology.

However, Iraq was unable clearly to improve the use of available technology, like other Arab countries, despite the availability of consulting firms and construction companies, millions of university graduates and about one million Arab engineers and hundreds of industrial companies and thousands of universities teachers.

The challenges facing Iraq, lies in the two groups, first resulting from major developmental problems, (food security, health, housing human rights, education, transport) and difficulty caused by the absence of the required scientific culture. The second is cultural in nature and include a special site independently. According to this scenario it requires the creation of systems of national technology take upon themselves the development of technology policies.

There are no clear technology policies in Iraq, but may be the presence of strategies development that include the lines of long-term development of science and technology indirectly that deals independently within the development process. In Iraq private enterprises will invest in technology (Higher Education ………… and others) in establishing a technological infrastructure.

The, international cooperation in Iraq is limited and weak at present, while the efficiencies gained by technological institutions in Iraq in coordination with existing competencies in the areas of social economic activity exists in society.

Under scientific siege, budgets reduced drastically, and technological institutions reduced also their expenses strongly which led to the non continuation of the previous level of innovative capacity, and the inability of the infrastructure and technology to do their core functions.

Iraq continued to practice, so far away from any genuine interest, the development of technology policy, due to many reasons already dealt with. It is worth mentioning here that any technology policies, in particular, depend on the quality of Iraq's exports. Petroleum Exporting Countries and Iraq, supplier of natural non-renewable sources reflected the policies of technology accordingly, it requires the decision- making process and technological identification of reliable channels between scientific community and the political leadership. Moreover, the wording of policy drafting requires determining the objective of technology that includes:

It is proposed that a variety of technology organizations should be related to the formulation of technology policies, including technological societies and other professional such as industrial and technological organizations.

These organizations have to operate in an environment different from the classical environments in recent years (currently confined to the production, dissemination and application of knowledge covering a range of disciplines and areas of application), but must evolve and interfere with a range of social and economic activities and cooperate with other private sector organization. Moreover, these organizations propose concepts and technological policy inputs.

Moreover, these technological organization should involve the largest possible number of workers participates in the preparation of technology policies, including representatives of government departments concerned economists, chambers of agriculture, industry, trade, non-governmental organizations and technological societies.

Proposal to develop a higher council for technology

Iraq's various institutions face difficulties in promoting technological capabilities and future planning of the technologic issues to acquire the capacity of innovation. This can not be achieved unless being done within the framework of science policy interrelated. Accordingly, it is required to use new methods of technology policies, and to develop integrated growth strategies that take into account of local and external technological conditions. To implement these, it is proposed to develop an academy or a higher council for technology. This Council (or academy) which is an institution with personal, moral, financial and administrative requirements characterized by autonomy should be linked to the Office of the Presidency.

The creation of the council facilitates and exercises technology policies, including matters related to the twenty-first century in Iraq. The need is to create some things that have been reviewed in the preceding paragraphs, including a brief analysis of technological capabilities, with a focus on research and development as well as the change that happened in technological institutions, leading to social and economic development.

Policies can be characterized by technological trends and are generally maintaining the balance between any movement toward the development of local technology and the establishment of links with external sources of technological knowledge that take place in several stages:

These include the input and output of technology policy with emphasis on the role of different institutions, including industrial, educational and private and scientific academy. It is also important to emphasize in this area on organizational matters that are between institution and how to submit proposals and receive guidance on technology policy that includes:

After gathering proposals for technology policy from various channels according to directions from the supreme authority, it is proposed, in this case, to adopt technological policy objectives (competition, trade barriers, the environment, new technologies). Then the policy with the top destinations should be approved by political leadership.

Thus, according to this scenario it is needed to start work for the establishment and completion of institutional arrangements and accordingly the following are accompanied the proposal.

- Evaluation and forecast of technological advances.
- Assess the market demand for technology.

- Planning and management of technology.

The technological policy decision should take into consideration the following:
- Be based on local capacity and technological progress in other places and to balance them.
- Contribution to regional and international programmes for the development of technology.

The role of the Supreme authority:
- It is proposed to take measures immediately to develop a policy and technological integration in social and economic development plans in Iraq, and the establishment of linkages with the private sector through specific proposals.
- Creation of private channels and specialized committees to prepare the requirement of technology policy and guide these channels with the supreme Council or academic parameters of this policy.

Development of national policies for Science and Technology
Science and Technology play a crucial role in shaping the challenges faced by individuals, organizations and nations constantly, (discoveries of genetic engineering, industrial human parties, mobile phone, biotechnology…etc.). The challenge that was facing by mankind at the beginning of the twentieth century is how all countries could benefit from the strength of science and technology.

However, Iraq was unable clearly to improve the use of available science and technology, like other Arab countries, despite the availability of consulting firms and construction companies, millions of university graduates and about one million Arab engineers and hundreds of industrial companies and thousands of universities teachers.

The challenges facing Iraq, lies in the two groups, first resulting from major developmental problems, food security, health, housing human rights, education, transport, and difficulty caused by the absence of the required scientific culture. The second is cultural in nature and includes a special site independently. According to this scenario it requires the creation of systems of national science and technology take upon themselves the development of science and technology policies.

Iraq made over the past decades considerable progress in several areas, and has increased the resources allocated to education, social services and infrastructure which has had a positive impact on the average per capita income and quality of life, followed by problems of concern (during the nineties). The per capita of the total GNP was decreased, the blockade halted efforts to diversify the economy and the adoption of the main sources of GNP on non- renewable mineral wealth.

There are some hypothesis for scientific and technological policies in the country which are limited in scope, effectiveness and ineffective strategies in the best of circumstances. So it was proposed to create new structures after some test on developed countries and developing countries.

The availability of financial resources in Iraq, especially before the nineties of the twentieth century has made great efforts in manufacturing, especially in the field of Military Industrialization. Iraq succeeded in building an independent military industrial base by enabling technology in modern manufacturing processes, however, strategies and manufacturing policies lacked effective action to develop local capacity and providing appropriate incentives for local people to be able in modern industrial technology.

Despite this, Iraq still needs to pursue innovative methods to meet the daunting challenges of a large number of other sectors of production and services. It also needs to develop scientific and technological capabilities of local traditional industries to modernize and to address a variety of social economic problems.

The scientific efficiency in Iraq itself has a capacity of innovative crucial role in facing challenges, in spite of standard conditions faced by the departure of many of them outside Iraq.

The limited scientific successes that have occurred in Iraq are the result of the efforts of some institutions of science and technology (Scientific Academy, the House of Alhikma, and institutions of higher education). On the other hand, important achievements were made in building the institutions mentioned, as well as in human resources development. But the institutions mentioned and others are still far from enabling their employees from playing a distinguished role in development. The linkages and synergies between the scientific and technological institutions of governmental organization, and the business world remain weak, despite the existence of some versions of contracting.

The spending on research and development in Iraq, at best, less than its counterpart in the Arab countries and more discouraging regarding the outputs of science and technology, to the degree that scientific and technological publications in specialized areas and the number of patents granted to institution and individuals are much less than the average of the corresponding figures in other developing countries.

As a result, the status of science and technology in Iraq needs a lot of attention. The input and output of scientific and technological point indicate the deficiencies in information networks, computers, advanced equipment, and scientific research.

These include, institutions (universities, research centers universities, atomic energy, etc.). Their shortcomings in Iraq are one of the reasons that led to the absence of scientific and technological policies, and the recognition of the limitations.

Are the challenges facing Iraq in the first two major problems arising from development (food security, health housing, human rights, education, transport) and the difficulty caused by the absence of the requisite political culture, and the second is cultural in nature, requiring the creation of national systems for science and technology take on a science policy and technology for Qatar.

That Iraq has acquired over the past decades significant progress in several areas, has increased resources allocated to education, social services and infrastructure, which had a positive impact on the average per capita income and quality of life, followed by such things into backwardness and closing (during the nineties) have per capita gross national product and stop the embargo has led to efforts to diversify the economy and the adoption of the main sources of gross national product on non- renewable mineral resources.

There are scientific and technological policies in Iraq, the premise of limited scope and effectiveness of strategies also ineffective in the best conditions. Therefore, it is considered a proposal for the development of new structures after the submission of some of the experiences of some developed and developing countries.

This can be explained as follows:

- The availability of financial resources in Iraq, especially before the nineties of the twentieth century reinforced the great efforts made in manufacturing, especially in manufacturing military. Iraq had succeeded in building an independent military industrial base by enabling technology in modern manufacturing processes, however, strategies and policies are free of manufacturing procedures effective way to develop local capacity and providing appropriate incentives for local people to be able to modern industrial technology.

- Iraq still needs to follow innovative methods to meet the daunting challenges of a large number of other sectors of production and services as well as to develop scientific and technological capabilities to update the local traditional industries and addressing a variety of social problems ..

- The scientific competencies in Iraq once upon innovative capacity has a crucial role in meeting the challenges in spite of the circumstances faced by the standard and leave many of them outside Iraq.

- The limited scientific successes that have occurred in Iraq are due to some institutions of science and technology (the Academy, the House of Wisdom, higher education institutions), these are important achievements in the areas of institution- building in question as well as in human resource development, the other is still far from to play an enabling role in development, and the linkages and interaction between the scientific and technological institutions- governmental and business world is still weak, although some versions of the contract between them.

- Expenditure on research and development in Iraq, at best, less than in Arab countries and the picture is more frustrated with the outputs of science and technology, so that science and technology literature in specialized areas and the number of patents granted to organizations and individuals much less than the average of the corresponding figures in developing countries other.

- As a result, the state of science and technology in Iraq needs a great deal of attention, input and output indices of scientific and technological points to deficiencies in information networks, computers, advanced equipment, scientific research.

- As a result of this fact it can determine the reasons as follows:
- The poor conditions in which the research is carried out.

- Lack of clarity in the criteria used for promotion careers.
- The case of destitution and suffering experienced by research institutions and educational institutions.
- Poor physical condition of the researchers.

Scientific and technological institutions in Iraq

These include institutions (universities, research centers, universities, educational institutions, the Atomic Energy… etc) and palaces in Iraq. The reasons that led to the absence of scientific and technological policies, the recognition of which is inadequate and ineffective management practices, the presence of structural deficiencies is a symbolic recognition of the need to develop their capabilities and should be in the next decade to work in a different environment to the environment of recent years.

These institutions are currently on:

- Production and dissemination of scientific research.
- Classification of Knowledge include a range of disciplines and areas of application.
- Training of some researchers.

And thus requires a complete reform and revitalize these institutions and to support and prioritize high- level and by increasing the competitiveness and compatibility interface and create an effective system of financing policies and link them to industry and social and economic activities.

That current levels of scientific and technological cooperation between Iraq and the Arab countries is very low as well as with foreign countries which are almost non- existent as evidenced by the lack of joint scientific research and the sum of joint publications. This refers to the need for Iraq to seek to strengthen their scientific and technological strengthening of scientific and technological interactions.

To strengthen communication and cooperation between the bid in Iraq with other Arab countries is a fundamental prerequisite in determining the scientific problems better and to obtain acceptable yields of knowledge available.

Among the basics of scientific cooperation to avoid turnkey contracts are void of any test of technology, which is almost a priority, involving scientists from two or more Arab countries and the increasing capabilities available to researchers in the country to attend scientific meetings and in more efficient use of international aid organizations.

The term technology transfer that technology to be acquired by learning through practice and not only through the transfer or importation of goods or technology services and therefore they are not the manufacture of pre-designed removable and caused the deficit technology.

The experience of Iraq in the transfer of technology are varied, including what has been achieved with foreign companies, which provided a comprehensive technological transactions and complex part of a strategy in the international market and the country has suffered from indiscrimination transfers that took place in the absence of a sound domestic policy to create a local independent in various fields technology.

Accordingly, Iraq is facing two problems the first concerning the search for modern technology, transport, and absorption, development and improvement and the second is related to technology development.

Science and technology policies in some countries of the world and United States

The United States adopted the U.S. policy of scientific and technological developments to:

- Addressing global challenges and the new National directly.
- The government impose a major role in assisting private companies to grow.
- Economic growth is the focus of policies:
  - The ability to competitiveness in the industry.
  - Creation jobs.
  - Creation an environment in technological innovation.
  - Coordination of technology between government departments.
  - A working partnerships between industry and the federal government and universities.

- Focus on new technology (information, communications, manufacturing technologies).
- To consider the basic science is the rule upon which all technological progress.

• Directing science policy toward the user.

Science and technology policies in the European Union

Represent the main targets of the following general framework for science and technology policies in Europe:

• Acting as a collective actor in the field of science and technology:
- Encourage the use of research facilities better.
- Confirming the international role of European research.
- Strengthening the scientific capacity and technology European countries.
- Development of the knowledge base.

• Enhancing the competitiveness of European industry on the international level.

• Narrow the gap between technical areas and disadvantaged areas in Europe.

• Meet the social and economic needs of the EU.

• Innovation with the participation of small and medium- sized enterprises.

• Development of human potential through the training of researchers.

The relevant authorities in the formulation of world policies and technology

Relevant authorities in the formulation of science and technology policies are:

• Proposed here to different destinations and a variety of institutions relevant in the formulation of science and technology policies, including higher education institutions, research and development institutions, the Academy, institutions, scientific societies and other professional organizations as well as for industrial, technological and educational.

• These institutions have to operate in an environment different from the environments in recent years not only being done on the production, dissemination and application of knowledge covering a range of assignments and areas of application, but can develop, interfere with a range of social and economic activities and cooperate with other private sector institutions. Furthermore it requires such institutions to propose the concepts of scientific and policy, inputs in accordance with a specific and include discussion.

- Therefore it should involve the largest possible number of relevant authorities in the preparation of science and technology policies, including the representative of government departments and professional economists, chambers of agriculture, industry, trade and non- governmental organizations and scientific societies.

Proposed development of a higher council for science and technology

Given the fact that Iraq and its institutions faced various difficulties in promoting scientific and technological capabilities and future planning of the issue of scientific and thus acquire the capacity to innovate only be achieved in a context of sound and coherent science policy.

It is supposed to do these new ways of science and technology policies and the development of an integrated growth strategies take into account the situation of domestic and external scientific and proposed for the implementation of the development of scientific academies or higher council for science and technology to do so.

This is characterized by the Council (or academic) being an institution with personal moral and financial and administrative independence linked to the Council of Ministers and shall achieve the following:

- Development of scientific and technology policies in accordance with the demands of current and future science.

- Actively contributing to the movement of scientific development, internal and external.

- To promote studies and scientific research in Iraq to keep pace with scientific progress in the world.

- The establishment of scientific links and close cooperation with the Arab and international destinations.

The Academy or the Supreme Council of:

- Members of the least number of 35 and not exceeding 40, including the President of the Council or the academy.
- Secretary- General.
- Require the member to be a scientist and a researcher in one of the branches of knowle4ge (agricultural, industrial, pure, medical, engineering) and have a broad access to more than a branch or branches of knowledge with a genuine product and the member shall be appointed by presidential decree and has particularly high in the hierarchy.
- The President of the Academy enjoy the rank of minister with one or more.
- The academy's secretary- general appointed a full- time among its members.
- Academy of various commissions specialist working in coordination with other scientific institutions in the country (Commission on Technology, Committee on Water, Energy Commission, Commission for information).
- As of the Academy of specialized services are also within the general framework of scientific knowledge (agricultural, medical, engineering, Pure).
- The Academy works to propose policies of science and technology and then brought to the presidency of the Republic for approval (after revision by its committees and Chambers).
- We are also working on the recommendation to grant material assistance to the centers, individuals and approval of the establishment of various scientific centers.
- Develop work contexts between scientific institutions and the stakeholders and the mechanism of cooperation between scientific institutions and relevant international institutions.

The Supreme Council for Technology the **Supreme**Council for Science and Technology:

- Council for Science and Technology
- Secretariat General
- Scientific bodies
- The executive

The Supreme Council for Science and Technology to achieve the following goals:

- Adoption of science and technology policies to serve the requirements of development plans.
- Contribute effectively to support the movement of scientific development and technology.
- Create an environment of scientific and technological discoveries to the creativity and stimulating the role of science and technology.
- Researchers and technologists in the public and private sectors provide forms of support and backing.
- The adoption of modern systems for the exchange of information between the institutions of science and technology in the country and work to develop mechanisms for coordination and cooperation between institutions.
- Strengthening the link States and Arab organizations, foreign and international and benefit from the products of research in the fields of science and technology to enhance the potential of relevant institutions in those areas in the country.
- Development the scientific and technological information for technological development.

Variety Council for Science, Technology and duties

Council consists of science and technology from a number of Ministers and the Secretary General of the Council on Science and Technology and the heads of the scientific bodies of the Council and others selected by the President of the Republic.

The Scientific Council of the obligations derived:

- Development trends and goals of the plans the development of science and technology.
- Consideration of the science and technology policies proposed by the scientific bodies and approval during the second semester of each year.
- Development of scientific bodies in accordance with priorities and requirements of the development of science and technology and national development plans.
- Choose a president and members of the scientific bodies.

- Decide on the proposals and recommendations from scientific bodies and the competent departments related to science and technology.
- Adoption of mechanisms of action within the formations of the Supreme Council for Science and Technology and issue instructions related to the implementation objectives of the council.
- Approve plans for the development of research and technological cadres inside and outside the country.
- Adoption of the percentages of how the distribution of resources available on the details of policies and plans proposed by the scientific bodies.

Secretariat General

- The Secretariat consists of a number of scientific bodies and the executive.
- Headed by a Secretary General appointed by the President of the Republic shall be the following duties:
    - Is due to the Council on Science and Technology.
    - Responsible for preparing the agendas of the Council for Science and Technology and the implementation of its decisions.
    - Supervisor of the proceedings of the scientific bodies and the executive.
    - Coordination with the concerned parties to implement the objectives of the council.

Scientific bodies

- Develop scientific bodies by the Technology Council, including, for example:
    - Scientific Committee for Pure Science.
    - Scientific Committee on Informatics.
    - Scientific Committee for Medical Sciences.
    - Scientific Committee for the Agricultural Sciences and Veterinary.
    - Scientific Committee for the Energy and Water.
    - Scientific Committee for the Engineering Sciences.
    - Scientific Committee for the Humanities.

- Scientific Committee for the Economic and Administrative Sciences.

• Scientific body, each headed by a member of the Council for Science and Technology.

• Board consists of a number of scientific experts selected by the Council for Science and Technology.

• Meets every scientific body once a month at least.

• The Director General is responsible the executive decision for a scientific body.

• Each scientific body has the following duties:

- Preparation of science and technology policies in the field of competence and updated annually during the first quarter of each year for presentation to the Council for Science and Technology.

- Trade- offs between the proposals for research projects for grant funding is available according to the priorities of science and technology policies and plans adopted by the pop- up and recommend their endorsement during the last quarter of each year.

- Recommendation of an award on the preparation of feasibility studies, technical and economic investment projects related to the development of human and material resources in the areas of science and technology under specialization.

The executive

• Linked with the Secretariat General a number of executive departments for the implementation and follow- up to the implementation of the resolutions of the Council of Science and Technology and decisions of the Secretary- General and scientific bodies and by the powers and competencies in the mechanisms of action adopted.

• Develop services derived:
- Executive Service first.
- Executive Service II.
- Executive Service III.
- Executive Service IV.
- Department of Foreign Relations.

- The department of information and documentation and dissemination of scientific and technical support.

- Head of each department of the units of the Secretariat staff at the Director General who is appointed by decree.

- The Director General of each of the executive decision of the Commission and one or more of the scientific bodies of the Council.

- Each executive service prepare the agenda for the relevant scientific work and the implementation of its decisions.

- Each executive service to announce details of science and technology policies relevant to its work during the third quarter of each year and determine the date that the research proposals and projects of scientific and technological institutions in the country.

- The Department for the operational contract for the implementation of projects and scientific and technological research and feasibility studies, technical and economic issues related to work and follow up on as scheduled.

- Specialized committees

    - Linked to the Authority the following specialized committees:
        - Commission for Basic and Applied Sciences.
        - Committee of Medical Sciences.
        - Committee of Agricultural Sciences and Veterinary Services.
        - Committee on Energy and Water.
        - Commission of new technologies.
        - Commission for Social and Human Sciences.
    - The number of specialists (at least six and not more than tens of part- time director nominated by the Board and the Board selected and appointed, including the presidency for two years, renewable once.
    - Each of the full- time Counselor is scheduled for administratively linked to the Director of the Commission.
    - The Council may create or abolish specialized committees and Hcesp need.

- Chambers
    - Associated with the following Council Chambers:
        - Department of Public bodies.
        - Service contracts and funding.
    - Head of each department employee the rank of director general.

- Directorates
  - Linked to the Authority three directorates:
    - Department of Administrative and Financial Affairs.
    - Department of Information and Documentation and publication.
    - Directorate of External Relations.
- Objectives of the Council on Science and Technology, the Council aims to achieve the following:
  - Identification of trends in national policies for science and technology.
  - Monitoring the movement of science and technology in the world and ensuring to keep in contact with the country and active and influential in it.
  - Promoting the movement of scientific and technological research in the country to serve the national development plans.
  - Creation a stimulating social environment for scientific and technological work.
  - Exploring the creativity, care and employment of the proceeds.
  - Scientists and researchers in all sectors provide help and support possible for them.
  - Development of mechanisms for coordination, collaboration and integration between institutions of science and technology in the country.
  - Strengthening the link state and the Arab and foreign organizations and international organizations in the field of science and technology.
- Functions of the Council on Science and Technology:
  - Adoption of science and technology policies proposed by the Commission on Science and Technology and its specialized during the second quarter of each year.
  - Creation specialized committee as needed.
  - Choose a president and members of specialized committees.
  - Action on the recommendations and proposals made by the science and technology.
  - Adoption of the mechanisms for cooperation and coordination between the Commission and institutions of science and technology in the country.

- Percentage of the allocation of resources available on the details of the plans of science and technology adopted.

- The functions of the science and technology: The Authority will achieve the goals of the Council and the implementation of the following tasks:

  - Preparation of science and technology policy and presented to the Council.
  - Recommend the introduction or cancellation of specialized committees and as needed.
  - Nomination of Chairman and members of specialized committees.
  - Preparation of mechanisms of cooperation and coordination between the Commission and institutions of science and technology in the country.
  - Leading the affairs of the Authority and the overall supervision of the work of specialized committees.
  - Participate in the elves country to cooperate with the Arab countries and foreign in the field of science and technology.
  - Develop mechanisms work between specialized committees and departments and directorates within the Authority.
  - Preparation of quarterly reports of progress and view the semi-annual reports to the Board as needed.

- The functions of specialized committees: Each of the specialized implementation of the following:

  - Preparation and paper science and technology policies in the area of competence and updated annually during the first quarter of each year.
  - View the proposal and the policy paper at a symposium attended by specialists, enlarged for the purpose of participating in the elaboration of policies, plans and is one of the basic principles of effective planning and review in light of the discussions in preparation for display.
  - Trade- offs between the proposals for projects and research received from the implementing agencies to compete for the available funding and adopting the recommendation during the last quarter of each year.
  - Recommendation of an award on the preparation of feasibility studies, technical and economic investment projects related to the

development potential of physical and human resources in the area of specialization.

- Functions of the Chambers

- The Department for the affairs of the following:

  ▪ Implementation of the resolutions of specialized committees after its ratification.

  ▪ Collection and compilation of project proposals and research in preparation for submission to the relevant thematic elves.

  ▪ Study and analysis of progress reports for projects and research contracted to be executed

- The Department for contracts and funding Mayati:

  ▪ To announce details of science and technology policies at the end of the second quarter of each year and determine the date of the project proposals and research and compiled for presentation to the specialized committees during the last quarter of each year.

  ▪ Contracts for the implementation of approved research projects and follow- up implementation.

  ▪ Gather and compile progress reports for projects and research-based and forwarded to the Department of Public bodies.

- Financing plans and activities of the style and rate:

- The resources and balancing the body of the following Financing plans and activities of the style and rate:

  ▪ Proportion (2%) of the profits of public companies in the state.

  ▪ Grant of central state budget.

  ▪ Allocations for the implementation of projects in a deliberate investment budget.

  ▪ Rent money brought about the outcome and publications.

  ▪ Donations sector companies mixed and private sectors.

- The Authority is spending on its activities under the special instructions.

- The Commission adopt the style of its grant- making to the party that contracted with the implementation of the specific activity and is given to those within the gates of the power exchange is not restricted to work but are subject to audit.

The Council of Science and Technology

Formation and duties

Council of science and technology consists of a number of ministers, secretary general of the science and technology and the chairpersons of the council and others selected by president of the ministers. It carries out the following duties:

- Working on a general trends, plans and objectives for the development of science and technology.
- Considering of science and technology policies proposed by scientific bodies and approved during the second quarter of each year.
- Developing scientific bodies in accordance with the priorities and requirements of the development of science, technology and national development plans.
- Selecting a president and members of scientific bodies.
- Deciding on the proposals and recommendations from scientific bodies and relevant competent departments of science and technology.
- Adopting working mechanisms within the council, and issuing instructions to implementation of its objectives.
- Adopting the development plans for cadres of research and technological work inside and outside the country.
- Adopting the percentage of how to distribute the resources available on the details of policies and plan proposed by scientific bodies.

- Goals of council on science and technology

The council aims to achieve the following:
- Identifying the trends in national policies for science and technology.
- Monitoring the movement of science and technology in the world and keep the country secure and in contact.
- Advancing the movement of scientific and technological research in the country to serve the national development plans.

- Creating a stimulating academic environment for scientific and technological work.
- Exploring creativity, nurturing and recruitment of dividends.
- Caring of scientists and researchers in all sectors and provide support and possible backing to them.
- Developing mechanisms for coordination, collaboration and integration between science and technology institutions in the country.
- Strengthening the links with the Arab and foreign countries and the international organizations, in science and technology.

• Functions of the council of science and technology:
- Adopting of science and technology policies proposed by the bodies on science and technology and its special committees during the second quarter of each year.
- Creating or canceling specialized committees and as needed.
- Selecting a president and members of specialized committees.
- Deciding on the recommendations and proposals from the body on science and technology.
- Adopting cooperation and coordination mechanisms between the body and science and technology institutions in the country.

• Functions of the science and technology body are represented by the achievement of the objectives of the council and implement of the following tasks:
- Preparation of science and technology policies presented to the council.
- Recommendation to create or cancel specialized committees and as needed.
- Nomination of the chairman and members of specialized committees.
- Preparation of cooperation and coordination mechanisms between the body and science and technology institutions in the country.
- Management of body affairs and the overall supervision of the work of the specialized committees.
- Participating in the country to cooperate with Arab and foreign countries in science and technology.
- Developing mechanisms between specialized committees and departments and divisions of the body.

-Preparation of quarterly reports to the progress of work and supply of semi-annual reports to the board as needed.

• Special functions of the committees

Each of the committees is supposed to carry out the implementation of the following:

-Preparing a science and technology policies in the field of competence and updated annually during the first quarter of each year.

-The presentation of a proposal of the policy paper at a symposium attended by specialists for the purpose of participating in the elaboration of policies plans, which is considered one of the basic principles of effective planning and then reviewed in the light of discussions in preparation for presentation.

-Differentiation between research and project proposals received from the actors to compete for available funding, and to recommend approval during the last quarter of each year.

-Recommendation of contracting for the preparation of technical and economic feasibility studies for investment projects relating to the development of human and material resources in the area of specialization.

• Function of departments

The department of public bodies carries out the followings:

-Implementation of the decisions of specialized committees after ratification.

-Collection and compilation of project proposals and research in preparation for submission to the relevant specialist committees.

-Analysis reports of the progress of work projects and research contractor to implement them.

The department of contracts and funding carries out the followings:

-Announcing the details of science and technology policies at the end of the second quarter of each year and determining the date of receipt of research project proposals, and compile them in preparation for presentation to the specialized commissions during the last quarter of each year.

-Contracting for the implementation of approved research projects, and monitoring implementation.

- Collecting and compiling reports of the progress of work and research projects approved and forwarded to the department of public bodies.

• Financing plans and activities and the methods of exchange

The resources and budget of the body from the following:

- Proportion of (2%) of the profits of the public companies in the state.
- Grant from the central state budget.
- Allocations to implement projects in a deliberate investment budget.
- Profits from the funds and publications.
- Donations of the corporations of the private sectors combined.

The body carries out the exchange on its activities under special instructions.

- The body adopts the method of providing the grant, for the contracted side to implement the specific activity and give the other side the power to work, but they are subjected to scrutiny and codification of the fundamentalist.

Figure (1)
General skeleton of a High Council of Science and Technology

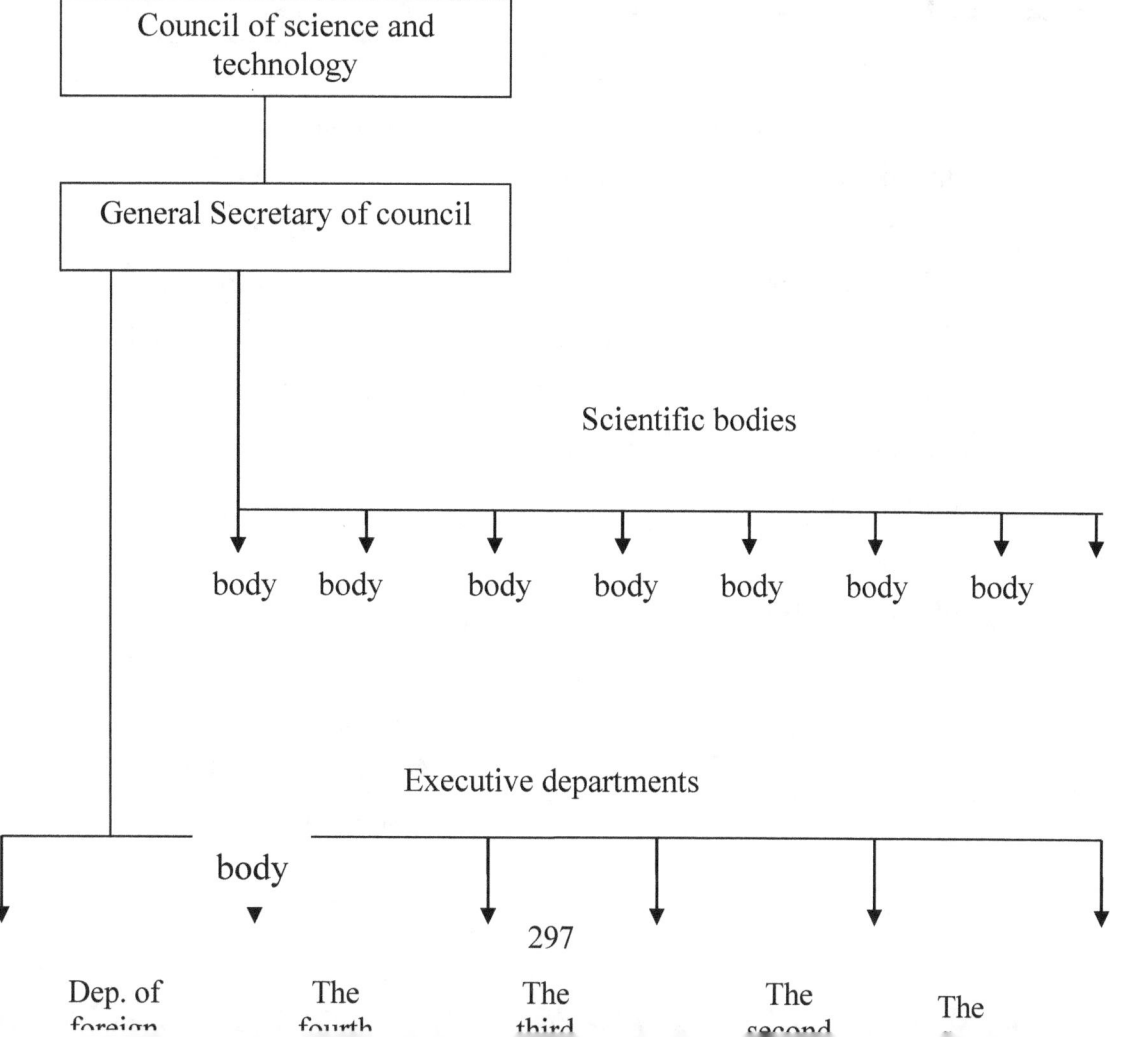

Figure (2)
Organizational skeleton of science and technology bodies

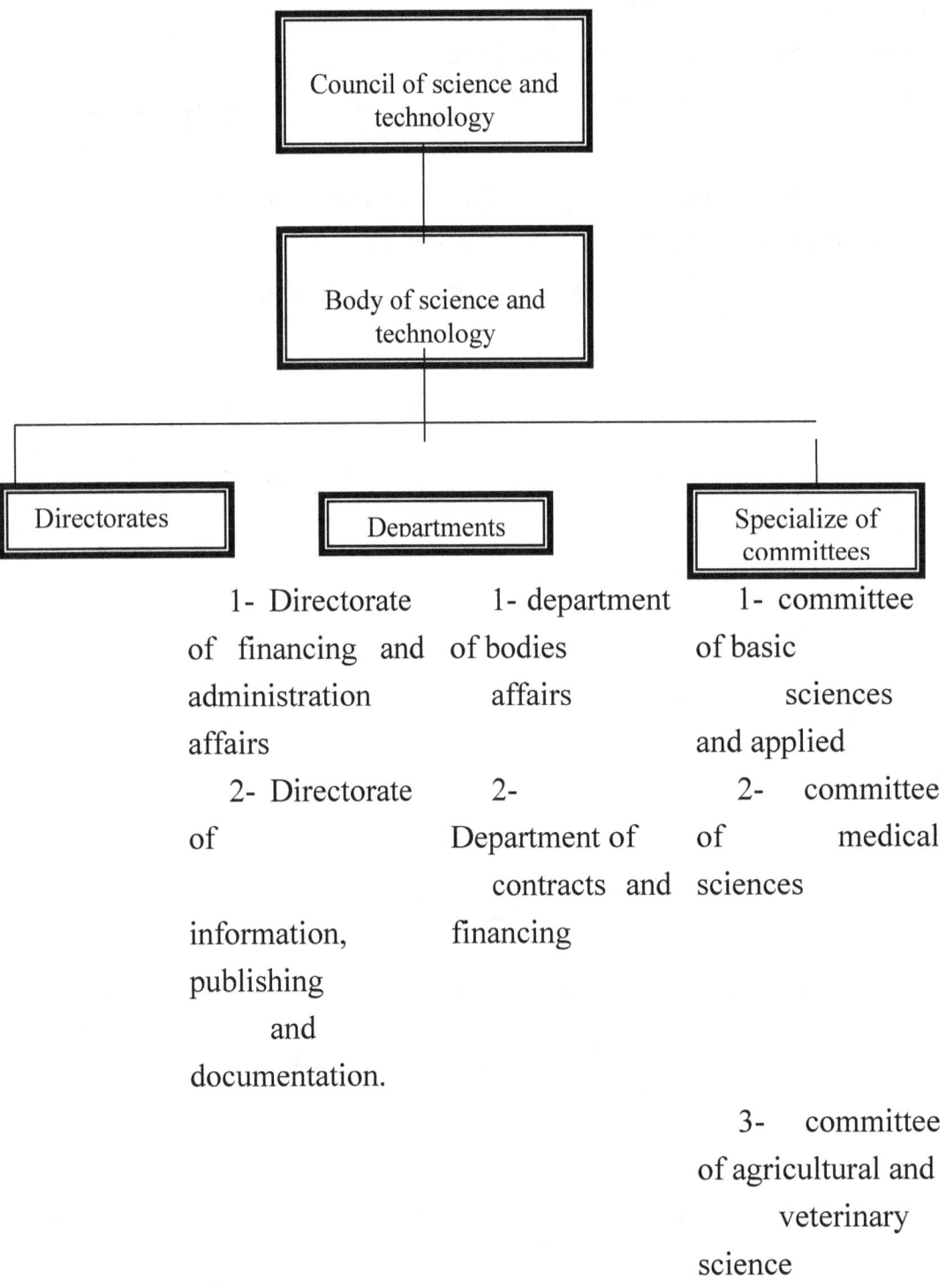

Scientific and technological cooperation *with Arab countries, foreign states and* international organizations

The current levels of scientific and technological cooperation between Iraq and Arab countries are very low, as well as with foreign countries which are almost non-existent as illustrated by the lack of joint scientific research and the outcome of joint publications. This refers to the need for Iraq in its quest to strengthen their scientific and technological interdependence.

To enhance communication and cooperation between Iraq with other Arab countries is a prerequisite in better determining the scientific problem and to obtain acceptable return of the available knowledge. Lounge of the basics of scientific cooperation is to avoid free turn key contains from any type of Technology, which is almost a priority involving scientists from two or more research fields, and incur the available possibilities to researchers to attend scientific meetings and make more use of international organizations.

The activity of technology transfer

The term technology transfer indicates that the technology is acquired through the transfer of goods or services technology and therefore is not equipped for transportation and technological deficits.

The experience of Iraq in the transfer of technology is varied and appropriate in a large proportion, completed with foreign companies which provided comprehensive and complex technological deals in the framework of international market strategy, and the country suffered from indiscrimination of transfer that took place in the absence of a sound domestic policy in various technological fields.

There are two problems facing Iraq, the first is concerning the search, transmission, absorption, development and improvement of modern technology, and second is related to technology development. According to that, Iraq needs in the area of technology transfer the following:

- The search for technological alternatives.

- The selection of appropriate technology.
- Adapting the selected technology.
- Identifying the problems of adapting modern technology.

Research and development "R & D" activity

The research and development are both vital and active in maintaining the quality of scientific personnel, in ensuring access to advanced science and promoting technology transfer. In addition it is provide early warning preparation of technological progress, industrial, agricultural and health, whereas its investment is guaranteed.

The efforts in research and development in Iraq may differ from what made similar efforts in other countries in terms of outcome, it is still inadequate to meet the challenges posed by scientific and technological developments and the process of globalization despite the allocation of funds required in this area, and the last twenty years is good evidence.

- Establishment of an information society.
- Innovation of small and medium-sized enterprises participation.
- Development of human potential through the training of researchers.

Science and technology policies in Iraq (Education and other sectors)

There are no clear scientific and technology policies in Iraq, but may be the presence of strategies development that include the lines of long-term development of science and technology indirectly that deals independently within the development process. In Iraq private enterprises will invest in science and technology (Education ………… and others) in establishing a scientific and technological infrastructure.

The scientific, international cooperation in Iraq is limited and weak at present, while the efficiencies gained by scientific institutions in Iraq in coordination with existing competencies in the areas of social economic activity exists in society.

Under scientific siege, budgets reduced drastically, and scientific institutions reduced also their expenses strongly which led to the non continuation of the previous level of innovative capacity, and the inability of the infrastructure and science and technology to do their core functions.

The formulation of science and technology policies

Iraq continued to practice, so far away from any genuine interest, the development of science and technology policy, due to many reasons already dealt with. It is worth mentioning here that any scientific and technology policies, in particular, depend on the quality of Iraq's exports. Petroleum Exporting Countries and Iraq, supplier of natural non-renewable sources reflected the policies of science and accordingly, it requires the decision-making process and scientific identification of reliable channels between scientific community and the political leadership. Moreover, the wording of policy drafting requires determining the objective of science and technology that includes:

- Promotion the growth of local companies specializing in services and manufacturing activities.
- Advancement of specialized institution.
- Increasing the movement of personnel in science and technology.
- Overlapping science policy and technology with a range of social and other economic activities.
- Development of institutions that will depend on science and technology that operates in a different environment.
- Strengthening the management structure of science and technology.
- Raising the level of resources.
- The effectiveness of R & D institutes.
- Development of technology transfer.
- Development of international cooperation.
- Improvement of researchers' wages and working conditions.
- The development of science and technology in long- term visions.
- The establishment needed in long term visions of all relevant institutions.

- Definition of the roles of effective government, private sector institutions, non- governmental organizations and professional associations.
- Adoption of progressive methods to build scientific capacity.
- Removal of duplication, overlapping and conflict in science organizations.
- Monitoring and evaluation and effective drafting in the relevant agencies in science and technology policies

It is proposed that a variety of science organizations should be related to the formulation of science and technology policies, including scientific societies and other professional such as industrial and technological organizations.

These organizations have to operate in an environment different from the classical environments in recent years (currently confined to the production, dissemination and application of knowledge covering a range of disciplines and areas of application), but must evolve and interfere with a range of social and economic activities and cooperate with other private sector organization. Moreover, these organizations propose concepts and scientific policy inputs.

Moreover, these scientific organization should involve the largest possible number of workers participates in the preparation of science and technology policies, including representatives of government departments concerned economists, chambers of agriculture, industry, trade, non-governmental organizations and scientific societies.

Industrial enterprise and technology

These institutions are required to have the presence of a specific set of targets linked with science policy and technology, which are consistent with a coherent vision for the future.

The responsibilities to the views of specialized advisory functions include characterization and strengthening the linkages and knowledge flows and providing technical services.

Proposal to develop a higher council for science and technology

Iraq's various institutions face difficulties in promoting scientific and technological capabilities and future planning of the scientific, issues to acquire the capacity of innovation. This can not be achieved unless being done within the framework of science policy interrelated. Accordingly, it is required to use new methods of science and technology policies, and to develop integrated growth strategies that take into account of local and external scientific conditions. To implement these, it is proposed to develop an academy or a higher council for science and technology. This Council (or academy) which is an institution with personal, moral, financial and administrative requirements characterized by autonomy should be linked to the Office of the Presidency.

The council carries out the followings:

- Development of science and technology policies in accordance with the demands of current and future science.
- Contributing to the scientific development of internal and external sources.
- Promotion of scientific studies and research in the country to keep abreast of scientific advances in the world.
- Establishment of scientific ties and close cooperation with Arab and international destinations.
-

The academy or the Supreme Council consists of:

- Members of not less than 35 not more than 40 including the President of the Council or the Academy.
- Secretary-General.
- The member should be a scientist and researcher in one of the branches of knowledge (agricultural, industrial, treatments, medical, engineering) and has a broad access to one or another branch of knowledge, and has genuine scientific product.

- The academy members should be appointed by the Prime Minister and enjoys the rank of Deputy Minister.
- The Academy full-time secretary-general is appointed from among its members.
- The Academy has a number multiple specialized committees working in coordination with other scientific institutions in the country (Commission on Technology, the Committee of Water, Energy Commission and Commission for information).
- The Academy has specialized offices within the framework of scientific knowledge (agricultural, medical, engineering, pure science).
- The Academy creates the suggestion of science and technology policies and then filed for the presidency for the Adoption (after being revised, by its committees and services).
- It also works with the recommendation to grant material assistance to the centers and individuals, adoption of the establishment of various scientific centers.
- Development of labor contexts between scientific institutions and the beneficiaries and the mechanism of cooperation between scientific institutions of Iraqi and international institutions involved.

Development of national policies for science and technology of Iraq

- The science and technology policy in Iraq
- Indicators and trends in science and technology policy in Iraq
  - Science and technology in Iraq
  - Scientific and technological institutions in Iraq
  - Scientific and technological institutions with Arab countries, foreign countries, international organizations
  - The activity of technology transfer
  - R&D activity
  - Science and technology policy in some countries of the world
- Evaluation of science and technology policies in the world and some Arab countries and foreign countries and international organizations

- Science and technology policy
- Science and technology policies in some Arab countries and foreign countries
- Science and technology policy in Iraq
  - The formulation of science and technology policy
  - The relevant authorities in the formulation of science and technology policy
  - Proposed development of a higher council for science and technology
  - The Supreme Council for Science and Technology
  - Composition of council for science and technology
  - Science and technology policy
- Policy for scientific research in educational institutions
- Models and technology policies and strategies in Iraq
- Biotechnology policy in Iraq
  - Development of biotechnology in Iraq
  - Biotechnology Future
  - A proposal for the development of biotechnology in Iraq

That science and technology play a crucial role in the formation of the challenges facing individuals, organizations and nations constantly, (discoveries of genetic engineering interestingly, human industrial, mobile, biotechnology) and the challenge facing mankind at the beginning of the twentieth century. But Iraq has not clearly improve the use of science and technology available to it, shared the Arab countries despite the availability of thousands of consulting firms, construction companies and millions of college graduates and about one million Arab engineer and hundreds of industrial companies.

Are the challenges facing Iraq in the first two major problems arising from development (food security, health housing, human rights, education, transport) and the difficulty caused by the absence of the requisite political culture, and the second is cultural in nature, requiring the creation of national systems for science and technology take on a science policy and technology for Qatar.

That Iraq has acquired over the past decades significant progress in several areas, has increased resources allocated to education, social services and infrastructure, which had a positive impact on the average per capita income and quality of life, followed by such things into backwardness and closing (during the nineties) have per capita gross national product and stop the embargo has led to efforts to diversify the economy and the adoption of the main sources of gross national product on non- renewable mineral resources.

There are scientific and technological policies in Iraq, the premise of limited scope and effectiveness of strategies also ineffective in the best conditions. Therefore, it is considered a proposal for the development of new structures after the submission of some of the experiences of some developed and developing countries.

This can be explained as follows:

• The availability of financial resources in Iraq, especially before the nineties of the twentieth century reinforced the great efforts made in manufacturing, especially in manufacturing military. Iraq had succeeded in building an independent military industrial base by enabling technology in modern manufacturing processes, however, strategies and policies are free of manufacturing procedures effective way to develop local capacity and providing appropriate incentives for local people to be able to modern industrial technology.

• Iraq still needs to follow innovative methods to meet the daunting challenges of a large number of other sectors of production and services as well as to develop scientific and technological capabilities to update the local traditional industries and addressing a variety of social problems .

• The scientific competencies in Iraq once upon innovative capacity has a crucial role in meeting the challenges in spite of the circumstances faced by the standard and leave many of them outside Iraq.

• The limited scientific successes that have occurred in Iraq are due to some institutions of science and technology (the Academy, the House of Wisdom, higher education institutions), these are important achievements in the areas of institution- building in question as well as in human resource development, the other is still far from to play an enabling role in development, and the linkages and interaction between the scientific and technological institutions- governmental and business world is still weak, although some versions of the contract between them.

- Expenditure on research and development in Iraq, at best, less than in Arab countries and the picture is more frustrated with the outputs of science and technology, so that science and technology literature in specialized areas and the number of patents granted to organizations and individuals much less than the average of the corresponding figures in developing countries other.

- As a result, the state of science and technology in Iraq needs a great deal of attention, input and output indices of scientific and technological points to deficiencies in information networks, computers, advanced equipment, scientific research.

- As a result of this fact it can determine the reasons as follows:
  - The poor conditions in which the research is carried out.
  - Lack of clarity in the criteria used for promotion careers.
  - The case of destitution and suffering experienced by research institutions and educational institutions.
  - Poor physical condition of the researchers.

These include institutions (universities, research centers, universities, educational institutions, the Atomic Energy… etc) and palaces in Iraq. The reasons that led to the absence of scientific and technological policies, the recognition of which is inadequate and ineffective management practices, the presence of structural deficiencies is a symbolic recognition of the need to develop their capabilities and should be in the next decade to work in a different environment to the environment of recent years.

These institutions are currently on:

- Production and dissemination of scientific research.
- Classification of Knowledge include a range of disciplines and areas of application.
- Training of some researchers.

And thus requires a complete reform and revitalize these institutions and to support and prioritize high- level and by increasing the competitiveness and compatibility interface and create an effective system of financing policies and link them to industry and social and economic activities.

That current levels of scientific and technological cooperation between Iraq and the Arab countries is very low as well as with foreign countries which are almost non- existent as evidenced by the lack of joint scientific research and the sum of joint publications. This refers to the need for Iraq to seek to strengthen their scientific and technological strengthening of scientific and technological interactions.

To strengthen communication and cooperation between the bid in Iraq with other Arab countries is a fundamental prerequisite in determining the scientific problems better and to obtain acceptable yields of knowledge available.

Among the basics of scientific cooperation to avoid turnkey contracts are void of any test of technology, which is almost a priority, involving scientists from two or more Arab countries and the increasing capabilities available to researchers in the country to attend scientific meetings and in more efficient use of international aid organizations.

The term technology transfer that technology to be acquired by learning through practice and not only through the transfer or importation of goods or technology services and therefore they are not the manufacture of pre-designed removable and caused the deficit technology.

The experience of Iraq in the transfer of technology are varied, including what has been achieved with foreign companies, which provided a comprehensive technological transactions and complex part of a strategy in the international market and the country has suffered from indiscrimination transfers that took place in the absence of a sound domestic policy to create a local independent in various fields technology.

Accordingly, Iraq is facing two problems the first concerning the search for modern technology, transport, and absorption, development and improvement and the second is related to technology development.

The research and development of strategic importance in maintaining the quality of scientific personnel in ensuring access to advanced science and promoting technology transfer. As well as providing an early warning system to prepare for technological change which is reflected on the consequences of technological progress and industrial, agricultural and health, it is in fact the investment is guaranteed.

That efforts in research and development in Iraq may be different on any progress made similar efforts in other countries in terms the challenges posed by scientific and technological developments and the globalization process.

And therefore the assumption that Iraq to take great interest in research and development in all areas as the rule of progress, and allocate proportions of a good national incomes for this purpose must be developed in the contribution of scientists and inventors and technicians as well as Iraqis and research centers in the state institutions and consulting offices in universities, keeping in mind that the benefits of R & D activities can not be performed without complex networks and relationships.

After reviewing the scientific and technological policies in Iraq and other countries, shows that they had found work by the institutions, centers independently. Iraq does not have the actual potential for the use of science and technology in the development process and this will help in assessing the effectiveness of scientific institutions and technology as well as create and revitalize national capacity to coordinate the contributions of those policies in the development and adopt the criteria as follows when evaluating science.

Science and technology policies in some countries of the world

United States

The United States adopted the U.S. policy of scientific and technological developments to:

- Addressing global challenges and the new National directly.
- The government impose a major role in assisting private companies to grow.
- Economic growth is the focus of policies:
  - The ability to competitiveness in the industry.
  - Creation jobs.
  - Creation an environment in technological innovation.
  - Coordination of technology between government departments.
  - A working partnerships between industry and the federal government and universities.
  - Focus on new technology (information, communications, manufacturing technologies).
  - To consider the basic science is the rule upon which all technological progress.
- Directing science policy toward the user.

The main targets of the following general framework for science and technology policies in Europe:

- Acting as a collective actor in the field of science and technology:
  - Encourage the use of research facilities better.
  - Confirming the international role of European research.

- Strengthening the scientific capacity and technology European countries.
- Development of the knowledge base.

• Enhancing the competitiveness of European industry on the international level.

• Narrow the gap between technical areas and disadvantaged areas in Europe.

• Meet the social and economic needs of the EU.

• Innovation with the participation of small and medium- sized enterprises.

• Development of human potential through the training of researchers.

Proposed development of a higher council for science and technology

Given the fact that Iraq and its institutions faced various difficulties in promoting scientific and technological capabilities and future planning of the issue of scientific and thus acquire the capacity to innovate only be achieved in a context of sound and coherent science policy.

It is supposed to do these new ways of science and technology policies and the development of an integrated growth strategies take into account the situation of domestic and external scientific and proposed for the implementation of the development of scientific academies or higher council for science and technology to do so.

This is characterized by the Council (or academic) being an institution with personal moral and financial and administrative independence linked to the Council of Ministers and shall achieve the following:

• Development of scientific and technology policies in accordance with the demands of current and future science.

• Actively contributing to the movement of scientific development, internal and external.

• To promote studies and scientific research in Iraq to keep pace with scientific progress in the world.

• The establishment of scientific links and close cooperation with the Arab and international destinations.

The Academy or the Supreme Council of:

- Members of the least number of 35 and not exceeding 40, including the President of the Council or the academy.
- Secretary- General.
- Require the member to be a scientist and a researcher in one of the branches of knowle4ge (agricultural, industrial, pure, medical, engineering) and have a broad access to more than a branch or branches of knowledge with a genuine product and the member shall be appointed by presidential decree and has particularly high in the hierarchy.
- The President of the Academy enjoy the rank of minister with one or more.
- The academy's secretary- general appointed a full- time among its members.
- Academy of various commissions specialist working in coordination with other scientific institutions in the country (Commission on Technology, Committee on Water, Energy Commission, Commission for information).
- As of the Academy of specialized services are also within the general framework of scientific knowledge (agricultural, medical, engineering, Pure).
- The Academy works to propose policies of science and technology and then brought to the presidency of the Republic for approval (after revision by its committees and Chambers).
- We are also working on the recommendation to grant material assistance to the centers, individuals and approval of the establishment of various scientific centers.
- Develop work contexts between scientific institutions and the stakeholders and the mechanism of cooperation between scientific institutions and relevant international institutions.

Council for Science and Technology:
- Council for Science and Technology
- Secretariat General
- Scientific bodies
- The executive

The Supreme Council for Science and Technology to achieve the following goals:
- Adoption of science and technology policies to serve the requirements of development plans.

- Contribute effectively to support the movement of scientific development and technology.
- Create an environment of scientific and technological discoveries to the creativity and stimulating the role of science and technology.
- Researchers and technologists in the public and private sectors provide forms of support and backing.
- The adoption of modern systems for the exchange of information between the institutions of science and technology in the country and work to develop mechanisms for coordination and cooperation between institutions.
- Strengthening the link States and Arab organizations, foreign and international and benefit from the products of research in the fields of science and technology to enhance the potential of relevant institutions in those areas in the country.
- Development the scientific and technological information for technological development.

Council consists of science and technology from a number of Ministers and the Secretary General of the Council on Science and Technology and the heads of the scientific bodies of the Council and others selected by the President of the Republic.

The Scientific Council of the obligations derived:

- Development trends and goals of the plans the development of science and technology.
- Consideration of the science and technology policies proposed by the scientific bodies and approval during the second semester of each year.
- Development of scientific bodies in accordance with priorities and requirements of the development of science and technology and national development plans.
- Choose a president and members of the scientific bodies.
- Decide on the proposals and recommendations from scientific bodies and the competent departments related to science and technology.
- Adoption of mechanisms of action within the formations of the Supreme Council for Science and Technology and issue instructions related to the implementation objectives of the council.
- Approve plans for the development of research and technological cadres inside and outside the country.

- Adoption of the percentages of how the distribution of resources available on the details of policies and plans proposed by the scientific bodies.
- The Secretariat consists of a number of scientific bodies and the executive.
- Headed by a Secretary General appointed by the President of the Republic shall be the following duties:
  - Is due to the Council on Science and Technology.
  - Responsible for preparing the agendas of the Council for Science and Technology and the implementation of its decisions.
  - Supervisor of the proceedings of the scientific bodies and the executive.
  - Coordination with the concerned parties to implement the objectives of the council.
- Develop scientific bodies by the Technology Council, including, for example:
  - Scientific Committee for Pure Science.
  - Scientific Committee on Informatics.
  - Scientific Committee for Medical Sciences.
  - Scientific Committee for the Agricultural Sciences and Veterinary.
  - Scientific Committee for the Energy and Water.
  - Scientific Committee for the Engineering Sciences.
  - Scientific Committee for the Humanities.
  - Scientific Committee for the Economic and Administrative Sciences.
- Scientific body, each headed by a member of the Council for Science and Technology.
- Board consists of a number of scientific experts selected by the Council for Science and Technology.
- Meets every scientific body once a month at least.
- The Director General is responsible the executive decision for a scientific body.
- Each scientific body has the following duties:

- Preparation of science and technology policies in the field of competence and updated annually during the first quarter of each year for presentation to the Council for Science and Technology.

- Trade- offs between the proposals for research projects for grant funding is available according to the priorities of science and technology policies and plans adopted by the pop- up and recommend their endorsement during the last quarter of each year.

- Recommendation of an award on the preparation of feasibility studies, technical and economic investment projects related to the development of human and material resources in the areas of science and technology under specialization.

The executive

- Linked with the Secretariat General a number of executive departments for the implementation and follow- up to the implementation of the resolutions of the Council of Science and Technology and decisions of the Secretary- General and scientific bodies and by the powers and competencies in the mechanisms of action adopted.

- Develop services derived:
- Executive Service first.
- Executive Service II.
- Executive Service III.
- Executive Service IV.
- Department of Foreign Relations.
- The department of information and documentation and dissemination of scientific and technical support.

- Head of each department of the units of the Secretariat staff at the Director General who is appointed by decree.

- The Director General of each of the executive decision of the Commission and one or more of the scientific bodies of the Council.

- Each executive service prepare the agenda for the relevant scientific work and the implementation of its decisions.

- Each executive service to announce details of science and technology policies relevant to its work during the third quarter of each year and determine the date that the research proposals and projects of scientific and technological institutions in the country.

- The Department for the operational contract for the implementation of projects and scientific and technological research and feasibility studies, technical and economic issues related to work and follow up on as scheduled.
- Specialized committees
  - Linked to the Authority the following specialized committees:
    - Commission for Basic and Applied Sciences.
    - Committee of Medical Sciences.
    - Committee of Agricultural Sciences and Veterinary Services.
    - Committee on Energy and Water.
    - Commission of new technologies.
    - Commission for Social and Human Sciences.
  - The number of specialists (at least six and not more than tens of part- time director nominated by the Board and the Board selected and appointed, including the presidency for two years, renewable once.
  - Each of the full- time Counselor is scheduled for administratively linked to the Director of the Commission.
  - The Council may create or abolish specialized committees and Hcesp need.
- Chambers
  - Associated with the following Council Chambers:
    - Department of Public bodies.
    - Service contracts and funding.
  - Head of each department employee the rank of director general.
- Directorates
  - Linked to the Authority three directorates:
    - Department of Administrative and Financial Affairs.
    - Department of Information and Documentation and publication.
    - Directorate of External Relations.
- Objectives of the Council on Science and Technology, the Council aims to achieve the following:
  - Identification of trends in national policies for science and technology.
  - Monitoring the movement of science and technology in the world and ensuring to keep in contact with the country and active and influential in it.

- Promoting the movement of scientific and technological research in the country to serve the national development plans.
- Creation a stimulating social environment for scientific and technological work.
- Exploring the creativity, care and employment of the proceeds.
- Scientists and researchers in all sectors provide help and support possible for them.
- Development of mechanisms for coordination, collaboration and integration between institutions of science and technology in the country.
- Strengthening the link state and the Arab and foreign organizations and international organizations in the field of science and technology.

- Functions of the Council on Science and Technology:

- Adoption of science and technology policies proposed by the Commission on Science and Technology and its specialized during the second quarter of each year.
- Creation specialized committee as needed.
- Choose a president and members of specialized committees.
- Action on the recommendations and proposals made by the science and technology.
- Adoption of the mechanisms for cooperation and coordination between the Commission and institutions of science and technology in the country.
- Percentage of the allocation of resources available on the details of the plans of science and technology adopted.

- The functions of the science and technology: The Authority will achieve the goals of the Council and the implementation of the following tasks:

- Preparation of science and technology policy and presented to the Council.
- Recommend the introduction or cancellation of specialized committees and as needed.
- Nomination of Chairman and members of specialized committees.
- Preparation of mechanisms of cooperation and coordination between the Commission and institutions of science and technology in the country.

- Leading the affairs of the Authority and the overall supervision of the work of specialized committees.
- Participate in the elves country to cooperate with the Arab countries and foreign in the field of science and technology.
- Develop mechanisms work between specialized committees and departments and directorates within the Authority.
- Preparation of quarterly reports of progress and view the semi-annual reports to the Board as needed.

• The functions of specialized committees: Each of the specialized implementation of the following:
- Preparation and paper science and technology policies in the area of competence and updated annually during the first quarter of each year.
- View the proposal and the policy paper at a symposium attended by specialists, enlarged for the purpose of participating in the elaboration of policies, plans and is one of the basic principles of effective planning and review in light of the discussions in preparation for display.
- Trade- offs between the proposals for projects and research received from the implementing agencies to compete for the available funding and adopting the recommendation during the last quarter of each year.
- Recommendation of an award on the preparation of feasibility studies, technical and economic investment projects related to the development potential of physical and human resources in the area of specialization.

• Functions of the Chambers
- The Department for the affairs of the following:
▪ Implementation of the resolutions of specialized committees after its ratification.
▪ Collection and compilation of project proposals and research in preparation for submission to the relevant thematic elves.
▪ Study and analysis of progress reports for projects and research contracted to be executed
- The Department for contracts and funding Mayati:
▪ To announce details of science and technology policies at the end of the second quarter of each year and determine the date of the project

proposals and research and compiled for presentation to the specialized committees during the last quarter of each year.

- Contracts for the implementation of approved research projects and follow- up implementation.
- Gather and compile progress reports for projects and research-based and forwarded to the Department of Public bodies.

• Financing plans and activities of the style and rate:

- The resources and balancing the body of the following Financing plans and activities of the style and rate:
  - Proportion (2%) of the profits of public companies in the state.
  - Grant of central state budget.
  - Allocations for the implementation of projects in a deliberate investment budget.
  - Rent money brought about the outcome and publications.
  - Donations sector companies mixed and private sectors.

- The Authority is spending on its activities under the special instructions.

- The Commission adopt the style of its grant- making to the party that contracted with the implementation of the specific activity and is given to those within the gates of the power exchange is not restricted to work but are subject to audit.

Establishing a database for research and development is done by monitoring the output of research and development of Iraqi universities and educational centres and measuring their interaction with industry and services sectors mechanisms:

• Each educational institution is developing a database on the output of scientific research in terms of numbers addresses research published and research abstracts presented at scientific conferences, and books published, and the number of researchers at the university level doctoral degree, master's, and patents.

• Each education institution number of scientific research and patents, and the Ministry of Education has established a network between data centers, educational institutions and building a central database in the ministry of scientific research in educational institutions and updated annually.

- Each education institution report on the envoys to a doctorate degree in the areas of knowledge.

- Each institution of higher learning inventory of available scientific potential of the library, and scientific periodicals, and information networks and scientific equipment.

- Output is provided by Education Research Mwsat turned on to the development of industry and services.

The role of science and technology in achieving the goals of educational policy in Iraq following areas:

- Universal primary education the expansion of democracy in education at all levels and stages coverage by using technologies such as methods of telecommunications and improve the school buildings.

- Adjusting the curriculum and development and continuously updated and linked to local needs.

- Adopting methods and teaching aids through the quality of the preparation of teachers and their working conditions and to provide assistance to teachers and educators and use media technology in the classroom.

- Adopting modern methods to provide assistance to the teacher of the burden of examinations and provides a means of measuring knowledge and achievement in a variety of content and capabilities.

- Provide the means of modern methods of planning, management, evaluation and follow- up and implementation at the school level.

Towards the formulation of policies of science and technology for the twenty- century atheist after the introduction of the Council, could have been exerted to develop policies, including science and technology related matters and the twenty- century atheist in Iraq, where the need to create some of the things that have been reviewed in the preceding paragraphs including a brief analysis of scientific and technological capabilities with a focus on research and development. As well as aspects of change that took place in institutions scientific in the country and the extent of participation of Arab scientists and technologists in the scientific activities and international technology, leading to achieve social and economic development.

Biotechnology policy in Iraq

Science and technology policy issue is not new, but it presents itself today sharply unprecedented because it has become crucial in the lives of peoples. Have taken many governments in developing policies of scientific and technological concepts of the adoption of a joint public again, especially from the degree of economic, social, political, and development of these policies, require confrontation and non- traditional and radical solutions to these premises. The preparation of scientific and technological policies of Iraq is one of the necessities of enhancing the national economy best use of natural and human resources through building on scientific and technological base solid and advanced. In accordance with this concept it proposes the formation of an institutional structure sets trends in national policies for science and technology monitor the movement in the world and believes that the country to keep in contact and have an active and influential. In which to achieve this must be linked to the institutional structure of the region of high expansive vision and has to absorb the philosophy of political and economic system and future outlook to achieve this end for this Foundation (FDA) to work through specialized committees characterized members have the competence and capacity in scientific and technological fields and have a broad spectrum of knowledge and vision with insight into the edges of science and the ability to devise and develop scientific attitudes and appropriate for the benefit of the country and the aspirations of these committees is proposed to form a committee on biotechnology that will prepare the policies in this area.

Development of biotechnology in Iraq

The application of biotechnology can be of far reaching impact in Iraq, which suffers from:

- An increase in the number of the population.
- A chronic shortage of food.
- Poor nutrition.
- Health.
- Environmental problems.

That biotechnology and genetic engineering of the areas where the country can achieve rapid progress and meaningful, particularly in access to food security and promote pharmaceutical industries.

There are various activities undertaken by various governments, academic institutions and non- governmental organizations in the fields of biotechnology and genetic engineering, especially in agriculture. And progress in the areas of biotechnology and genetic engineering highlights the importance of investment in basic sciences which form the backbone of continued progress in science and technology, mainly because of the height of the research and development in biotechnology in the country. And the importance of organizing the sequence of the human genome, an event that can match a human landing on the moon, and is described as a landmark in the history of science will serve to maximize the research in human biology which focuses on the spread of diseases such as cancer. The Universal Declaration on Human genome and human rights adopted by the General Conference of UNESCO in 1997 and the major breakthroughs in molecular Biology and genetic engineering that have raised many legal issues and ethical and social development.

The general principles for the development of biotechnology in Iraq

- Definition the objectives of national development in the fields of science and technology, especially biotechnology.
- Definition of biotechnology strategy in Iraq combined with the proposed policies for science and technology.
- The introduction of technology outreach programs particularly vital at various stages of scientific education.
- The introduction of appropriate legislation to promote the aspects of the pharmaceutical industry.
- Creation linkages and partnerships with Arab and Islamic countries in the areas of biotechnology and genetic engineering in order to facilitate cooperation between these countries by governments, industry and academic staff.
- Continuation to deal with developments in basic science and non-marginalization of the backbone of the development of science and technology.

The principles of this development is represented:

- Avoid the use of any technology that could lead to finding unexpected material that can be harmful to health.
- It is obligatory to avoid the exploitation of any technology that could have environmental effects and can not be repaired, but the proof that the products did not cause any significant damage to the environment.
- It is inappropriate to put people and the environment even less degrees of risks to note that the current products of genetically engineered not of little value.
- It is not right to justify the exploitation of technology of high-risk because of the principle on the basis of sound scientific grounds that this principle could result in useful products in the future.
- The application of biotechnology to agriculture must be based on a scientific basis away from the trends of foreign trade (GMOs).
- Permission to postpone the deployment of genetically modified and the use of genetically engineered food to be a degree of knowledge which enables to judge the safety that can bring to human health and the environment in order to utilize this technology.

The modalities proposed for the development of biotechnology can be described as follows:

- Encourage and support biotechnology in Iraq, pharmaceutical and industrial projects that have bases in the Arab and Islamic countries.
- Development of databases of human resource, especially related to biotechnology and genetic engineering in the country to facilitate the assessment of the strengths and weaknesses of the national.
- Support research and development based on a confluence of branches of knowledge in various fields related to biotechnology and to ensure that development required human resources.
- Provide money and government support for medical diagnostic applications in addition to the therapeutic as well as biotechnology and gene therapy.
- Encourage and support the deployment of research materials of good quality.
- Establish a fund for biotechnology in order to transfer technical expertise from other countries.

Biotechnology Future

When studying the development of technology relating to food and health to meet the basic humanitarian needs in Iraq, it is the subject of food security in the country rank high on the concerns and increase the importance and the intensity of gravity of the subject, keeping in mind the blockade and the lack of renewable resources and the deterioration of productivity, and technical backwardness as well as other conditions.

The need for the application of biotechnology to meet the need for basic food and health is important and possible to review some of these things, especially those related to nutrition vital. Example, there is the development of biotechnology applied to livestock production and those related to face shortages of food, where it took the last frighteningly worse and more tragic. And the food crisis in the country threaten the security and future of the people and can not be met only with hard work and a plan to sharply improve the means of agricultural production.

Biotechnology will play a significant role in this area and develop a plan and can be summarized components of the plan proposed in the biotechnology, which consists of:

- Propagating minute through agriculture textiles.
- Refers to the genome and molecular characterization of all living species.
- Bioinformatics and includes collection of data from genomic analysis and compilation of forms, it is easy and the plans for the future of biotechnology can be formulated in several priorities:

- Food security.
- Increase and improve agricultural production and breeding higher food production as well as access to the kinds of resistance to diseases and pests and protection of plant genetic diversity.
- Production of pharmaceuticals from plant material biologically active.
- Production of vaccines and antibody monoclonal.
- The use and frequency of agricultural products to produce ethanol and acetone and Albiotanol.

Food Security

Characterized by agricultural biotechnology applications in the future Iraq as being promising for the purpose of providing the requirements for agricultural production and food security and sustainability, such as stress resistance and non- living (drought and salinity and to provide options for rotation of my life better and to conserve natural resources. Iraq is not the location of the use of genetically modified products that many of the advanced techniques are not clear in the fields and the mentality of farmers, but can be expected in the future for crop improvement in methods of accurate and fast. The use of genetics to address the functional characteristics of complex helps to maintain the genes and improve the quality of food and natural resources management using Adaut Almagbp efficient and Iraq is supposed to be an active participant in this area in order to obtain the essential needs of food security through research of atopic being done in collaboration with bodies international, regional and in particular the use of biotechnology in food security.

Agricultural production

Despite the importance of industry and health, the proposed priorities must take into account the agricultural biotechnology for two main reasons:

- The research conducted on plants to improve the crop is directly related to the specific prevailing environmental conditions, while those related to health and the industry is more difficult in the country.

- Preliminary data indicate that most of the research activities in the country related to agriculture.

The following themes of agricultural biotechnology could be suggested for use in Iraq in the future:

- Gene transfer techniques which Twfr of transgenic plants resistant to many of the nurses and insects (insect disease), insecticides as well as degrees of resistance to heat, drought and salinity.

- Techniques for control of natural resources and the environment.

The additional proposals for plants included:

Plant Biotechnology

- Develop an institution for bioremediation of sewage and water use in the treatment of those in agriculture and to identify agricultural sites.

- Develop a means of production of vaccines using methods and efficient use of agricultural waste such as sugar beet molasses and straw of rice and maize.

- To conduct research on the use of bio fertilizer production to increase rice production in Iraq due to the impact of fertilizers vital hospitality great when used as an alternative to some types of fertilizers because of its significant benefits from the most important to reduce severe pollution of groundwater as a result use of chemical fertilizers.
- Use methods of biotechnology and the improvement and development of insecticides to control insects plant the main alternative to chemical pesticides in order to avoid the risk, but recent work on the production of pesticides and devices require special materials.
- Increasing the protein content of rice crop, which is characterized by being a choice for the Iraqi people and the most important crops in Iraq and the most scalable in the unique conditions of stress.
- Handouts developed resistance to disease and the application of tissue culture to improve and compensate for the shortfall output and the large number of palm trees.
- Conduct future research to overcome the difficulties related to the early prosperity and lack of consistency plants reproduced.
- Production of secondary materials using tissue culture for example, used in medical, industrial and other lines by testing the plant to withstand cellular stress on salt and drought, as well as the production of potato virus-free, as well as the exact Propagating plants.

Animal Biotechnology

The introduction of biotechnology for movement of embryos in cattle, fish, for example the adoption of genetic improvement of cultivated domestic animal breeds there is a need for Qatar's importance in the propagation and the election of the number of animals with high standards of quality in production and can avoid the disease.

Microbial Biotechnology

Biotechnology refers to:

- Microbial biotechnology for the production of ethanol by products of sugar and the production of methanol from industrial and agricultural waste.
- Microbial genetics and their use in removing, destroying and conversion of pollutants to the installers of nitrogen and the formation of yeast isolates from the "S. cervisiae" are for the consumption of cellulose or lactose.

- Appropriate use of biotechnology to convert biomass into biofuel and biogas and biomass conversion to agricultural and bio-fuel and fertilizer adoption of various wastes for the production of biomass and a rice crust potential applications.

- Bio-conversion of waste cellulose followed by the production of protein-rich microbial biomass of the product which decompose water.

- The use of microbial treatment to remove oil and metal chromium in the process of or remove the ammonia.

Technology vital health

It requires technical vital health in Iraq, in general great efforts to bridge the gap between developed and diagonals in the country to obtain the desired goal in health care.

- Establishment a center for bone marrow transplant in a hospital provides a number of specialists as well as equipment and materials.

- Start research projects in the field of gene manipulation in the tumors and other genetic diseases.

- Expansion of the pharmaceutical industry in the country and its development to meet local requirements as well as the modalities for the use of biotechnology.

Environment

The environment:

- Use of natural organisms (yeast, fungi and plants) to convert hazardous materials in the soil.

- Use of persistent micro-organism in the drainage systems for sewage and industrial sites as well as cleansing.

- Use of biotechnology to avoid contamination from the use of bioreactors for the products of hazardous materials.

- Forensic applications technology to forensic science.

Cooperation with international agencies

These include:

- Cooperation with the Islamic and international agencies that are required in the present time.

- Requiring trained scholars of Islamic and Arab countries that are directly involved in testing and the transfer of various biotechnologies.

- Development of short training courses for postgraduate students international organizations such as the UN agency of Educational and Cultural Organization (United Nations Educational Scientific and Cultural Education and UNESCO).
- Training of health personnel, especially doctors in the field of bone marrow transplantation and to assist in gene therapy, especially in cases of cancer of the blood and lymph gland tumor.
- Broadening the base of biotechnology such as the need to characterize the collection and preservation available mainly in the gene bank globally and provide information on all genes.
- Developing training programs for joint cooperation with various countries in the following fields in the area of biological control:
  - The exchange of biological information.
  - Mass production of the insect host.
  - Put out a vital crop for insects and the development of the joint projects.
  - The computerization of information and the development organization to link networks in different countries.
  - Training in the different features of the control of the House.

Foundation Center for Biotechnology in the light of Islamic treads provided and can be concluded that the status of biotechnology in the Islamic countries has become a necessity, especially those with deeper experience in biotechnology, which can greatly help in establishing such a center.

The application of modern genetics, genetic engineering fundamentals of moving and developing improvements to allow farmers to produce new varieties and in various forms more quickly and at lower cost and in contrast to traditional methods that require time, effort. The followings are some applications of genetic engineering in plant improvement:

- Improving the quality of food, especially protein stored and other materials.
- The production of plants capable of nitrogen fixation and private non-legumes.
- Production of new crops resistant to:
  - Pesticides.
  - Drought and therefore can be grown in deserts.

- Various pests and plant diseases
- Salinity and therefore can be grown in sea water.
- Frost in the country and in the cold winter nights.

- The transfer of genes between plants for the production of new types, especially the transfer of genes between species long descent through the development of new varieties.
- The production of plants with great ability in the operations photosynthesis.
- Cultivation of plant tissue in the hybrid types of long descent and even on the cellular mating in different ways.
- A plant products by micro-organisms through the transfer of plant genes and bring them into micro-organisms such as bacteria.
- The production of bio-pesticides "Production of Biocides".
- The production of bio fertilizers, "Production of Bio fertilizers".
- Tissue Culture "Tissue Culture Techniques".
  - Production of potato tubers and seedlings free of viral diseases.
  - Mass production of palm dates.
  - Production of the fruit crop.
  - Production of secondary materials.

Applications in animal

The progress that has been associated with new technologies such as artificial insemination and the use of female incubators and other techniques, especially genetic engineering to improve the offspring of animals through the production of useful new strains characterized by the following:

- Increase the number of resulting offspring.
- To contain higher amounts of red meat and with less fat.
- Its capacity for generating a great deal of milk with respect to types of livestock.
- Considerable weight and an increase in the number of eggs produced with respect to poultry.
- Specifications for the excellent wool sheep.

May contribute to genetic engineering in the area of changes in some animals in the following specifications:

- The possibility of transferring nuclei from one object to another.

- The use of livestock as a factory for the production of certain hormones and proteins that can be counted in milk and through:
  - Production of animals that carry transgenes known animals.
  - Either way, the other used to introduce genes into animals the use of viral vector known as the virus can be cultured bounced inside the desired gene is then inserted into the virus in animal cells to merge with the animal gene and begins to express itself within the animal.
  - Can take advantage of this method to introduce genes with essential medical value to humans.
  - The creation of a hormone in the cells of the individual in the dairy cattle only.
  - Production of some clotting factors (factor No. 9) responsible for the hemorrhagic diseases.
  - The production of many drugs such as interleukin-2 (to treat some types of cancer) in the rabbit.
  - Production of human growth hormone.
- Increased livestock productivity through:
  - Animal hormone injections.
  - The introduction of genes to increase the amount of hormones.
  - Increase the amount of wool.
  - Reduce the amount of fat.
- Building genetic maps of livestock, poultry and fish.
- Purification plant embryos in cattle.

Some applications in the industry

Of the most important applications in the industry to use objects allocated for Genetic Engineering to perform the following tasks:

- The production of antibiotics (the amount of larger, better quality, lower cost).
- The production of enzymes or different drugs that have medicinal value.
- Pollution control oil during the cracking of chemical compounds or remove pollution.
- Convert human waste to the food of a high content of protein for use in animal feed.

- Manufacturing and production of materials for energy such as ethanol, methanol and acetone.

In the treatment of oil- contaminated soil can develop a technology of high efficiency to help the oil sector to improve its products and expand, markets and increase prices and features of this technology as a clean environment and ensure the safety of its implementation contribute to economic return.

The project includes:

- To intensify and isolate bacterial strains that have the ability to remove sulfur.
- Increase the capacity of bacteria to remove sulfur and are counted at this stage, genetic engineering techniques, where is the transfer of specific genes carrying the desired qualities and planting, fast- growing bacteria plasmids.

Applications in humans

The study of genetic diseases is due to change in chromosome number or change in composition or changes at the molecular level of the gene, and is recognized as a gene that causes the disease gene through:

- The family.
- Use of modern techniques.
- Medical consultations: The family is a platform for diagnostic tests.
- Survey and genetic diagnosis of genetic diseases by:
- Ultrasound.
- Identify the presence of chromosomal changes or chemical weapons.
- New technologies
- Diagnosis and presence of specific genes in pre- reproductive individuals (which may be an individual carrier of the disease as a result of a new gene without the appearance of signs of illness it).
- The presence of the gene in the fetus before birth.
- Development of a sophisticated system for the diagnosis.
- Evolution of techniques of PCR.
- Develop the use of genetically modified animals in medical research.

Environment

- The use of vital signs "Bio-indicators" and controls the vital "Biomonitors" to detect the levels of pollution.
- Cracker diversity and the recycling of industrial waste and agricultural products.
- Development of resistant plants.

Other areas of research

- Technology enzymes
  - Enzymes for the diagnosis.
  - Clinical biochemistry.
  - Quality control of food.
  - Industrial processes.
  - Environmental control.
- Industrial enzymes
  - Design and analysis of bioreactors.
  - Waste water treatment.
  - The technology of the vaccine.

Transfer of biotechnology

The use of an information base for the diagnosis of "STR" (Short Tandem Repeats).

- Design prefixes "Primers Design".
- PCR amplification for the purpose of the STR.
- Analysis of the STR.
- PCR amplification of the mind "(SNPs) Single Nucleotide Polymorphism".
- A diagnosis of SNPs enzymes.
- A diagnosis of SNPs, respectively.
- PCR amplification.

Genetic fingerprint "Finger Printing

Models of technology policies and strategies in Iraq

- Biotechnology policy in Iraq
  - Development of biotechnology in Iraq
  - Biotechnology Future

- A proposal for the development of biotechnology in Iraq

Development of biotechnology in Iraq

The application of biotechnology can be of far reaching impact in Iraq, which suffers from:
- An increase in the number of the population.
- A chronic shortage of food.
- Poor nutrition.
- Health.
- Environmental problems.

That biotechnology and genetic engineering of the areas where the country can achieve rapid progress and meaningful, particularly in access to food security and promote pharmaceutical industries.

There are various activities undertaken by various governments, academic institutions and non- governmental organizations in the fields of biotechnology and genetic engineering, especially in agriculture. And progress in the areas of biotechnology and genetic engineering highlights the importance of investment in basic sciences which form the backbone of continued progress in science and technology, mainly because of the height of the research and development in biotechnology in the country. And the importance of organizing the sequence of the human genome, an event that can match a human landing on the moon, and is described as a landmark in the history of science will serve to maximize the research in human biology which focuses on the spread of diseases such as cancer. The Universal Declaration on Human genome and human rights adopted by the General Conference of UNESCO in 1997 and the major breakthroughs in molecular Biology and genetic engineering that have raised many legal issues and ethical and social development.

The general principles for the development of biotechnology in Iraq
- Definition the objectives of national development in the fields of science and technology, especially biotechnology.

- Definition of biotechnology strategy in Iraq combined with the proposed policies for science and technology.
- The introduction of technology outreach programs particularly vital at various stages of scientific education.
- The introduction of appropriate legislation to promote the aspects of the pharmaceutical industry.
- Creation linkages and partnerships with Arab and Islamic countries in the areas of biotechnology and genetic engineering in order to facilitate cooperation between these countries by governments, industry and academic staff.
- Continuation to deal with developments in basic science and non-marginalization of the backbone of the development of science and technology.

The principles of this development is represented:
- Avoid the use of any technology that could lead to finding unexpected material that can be harmful to health.
- It is obligatory to avoid the exploitation of any technology that could have environmental effects and can not be repaired, but the proof that the products did not cause any significant damage to the environment.
- It is inappropriate to put people and the environment even less degrees of risks to note that the current products of genetically engineered not of little value.
- It is not right to justify the exploitation of technology of high- risk because of the principle on the basis of sound scientific grounds that this principle could result in useful products in the future.
- The application of biotechnology to agriculture must be based on a scientific basis away from the trends of foreign trade (GMOs).
- Permission to postpone the deployment of genetically modified and the use of genetically engineered food to be a degree of knowledge which enables to judge the safety that can bring to human health and the environment in order to utilize this technology.

The modalities proposed for the development of biotechnology can be described as follows:

- Encourage and support biotechnology in Iraq, pharmaceutical and industrial projects that have bases in the Arab and Islamic countries.
- Development of databases of human resource, especially related to biotechnology and genetic engineering in the country to facilitate the assessment of the strengths and weaknesses of the national.
- Support research and development based on a confluence of branches of knowledge in various fields related to biotechnology and to ensure that development required human resources.
- Provide money and government support for medical diagnostic applications in addition to the therapeutic as well as biotechnology and gene therapy.
- Encourage and support the deployment of research materials of good quality.
- Establish a fund for biotechnology in order to transfer technical expertise from other countries.

Biotechnology Future

When studying the development of technology relating to food and health to meet the basic humanitarian needs in Iraq, it is the subject of food security in the country rank high on the concerns and increase the importance and the intensity of gravity of the subject, keeping in mind the blockade and the lack of renewable resources and the deterioration of productivity, and technical backwardness as well as other conditions.

The need for the application of biotechnology to meet the need for basic food and health is important and possible to review some of these things, especially those related to nutrition vital. Example, there is the development of biotechnology applied to livestock production and those related to face shortages of food, where it took the last frighteningly worse and more tragic. And the food crisis in the country threaten the security and future of the people and can not be met only with hard work and a plan to sharply improve the means of agricultural production.

Biotechnology will play a significant role in this area and develop a plan and can be summarized components of the plan proposed in the biotechnology, which consists of:

- Propagating minute through agriculture textiles.

- Refers to the genome and molecular characterization of all living species.
- Bioinformatics and includes collection of data from genomic analysis and compilation of forms, it is easy and the plans for the future of biotechnology can be formulated in several priorities:

   - Food security.
   - Increase and improve agricultural production and breeding higher food production as well as access to the kinds of resistance to diseases and pests and protection of plant genetic diversity.
   - Production of pharmaceuticals from plant material biologically active.
   - Production of vaccines and antibody monoclonal.
   - The use and frequency of agricultural products to produce ethanol and acetone and Albiotanol.

Food Security

Characterized by agricultural biotechnology applications in the future Iraq as being promising for the purpose of providing the requirements for agricultural production and food security and sustainability, such as stress resistance and non- living (drought and salinity and to provide options for rotation of my life better and to conserve natural resources. Iraq is not the location of the use of genetically modified products that many of the advanced techniques are not clear in the fields and the mentality of farmers, but can be expected in the future for crop improvement in methods of accurate and fast. The use of genetics to address the functional characteristics of complex helps to maintain the genes and improve the quality of food and natural resources management using Adaut Almagbp efficient and Iraq is supposed to be an active participant in this area in order to obtain the essential needs of food security through research of atopic being done in collaboration with bodies international, regional and in particular the use of biotechnology in food security.

Agricultural production

Despite the importance of industry and health, the proposed priorities must take into account the agricultural biotechnology for two main reasons:

- The research conducted on plants to improve the crop is directly related to the specific prevailing environmental conditions, while those related to health and the industry is more difficult in the country.
- Preliminary data indicate that most of the research activities in the country related to agriculture.

The following themes of agricultural biotechnology could be suggested for use in Iraq in the future:
- Gene transfer techniques which Twfr of transgenic plants resistant to many of the nurses and insects (insect disease), insecticides as well as degrees of resistance to heat, drought and salinity.
- Techniques for control of natural resources and the environment.

The additional proposals for plants included:

Plant Biotechnology
- Develop an institution for bioremediation of sewage and water use in the treatment of those in agriculture and to identify agricultural sites.
- Develop a means of production of vaccines using methods and efficient use of agricultural waste such as sugar beet molasses and straw of rice and maize.
- To conduct research on the use of bio fertilizer production to increase rice production in Iraq due to the impact of fertilizers vital hospitality great when used as an alternative to some types of fertilizers because of its significant benefits from the most important to reduce severe pollution of groundwater as a result use of chemical fertilizers.
- Use methods of biotechnology and the improvement and development of insecticides to control insects plant the main alternative to chemical pesticides in order to avoid the risk, but recent work on the production of pesticides and devices require special materials.
- Increasing the protein content of rice crop, which is characterized by being a choice for the Iraqi people and the most important crops in Iraq and the most scalable in the unique conditions of stress.

- Handouts developed resistance to disease and the application of tissue culture to improve and compensate for the shortfall output and the large number of palm trees.
- Conduct future research to overcome the difficulties related to the early prosperity and lack of consistency plants reproduced.
- Production of secondary materials using tissue culture Kalgulwyat for example, used in medical, industrial and other lines by testing the plant to withstand cellular stress on salt and drought, as well as the production of potato virus- free, as well as the exact Propagating plants.

Animal Biotechnology

The introduction of biotechnology for movement of embryos in cattle, fish, for example the adoption of genetic improvement of cultivated domestic animal breeds there is a need for Qatar's importance in the propagation and the election of the number of animals with high standards of quality in production and can avoid the disease.

Microbial Biotechnology

Biotechnology refers to:
- Microbial biotechnology for the production of ethanol by products of sugar and the production of methanol from industrial and agricultural waste.
- Microbial genetics and their use in removing, destroying and conversion of pollutants to the installers of nitrogen and the formation of yeast isolates from the "S. cervisiae" are for the consumption of cellulose or lactose.
- Appropriate use of biotechnology to convert biomass into biofuel and biogas and biomass conversion to agricultural and bio-fuel and fertilizer adoption of various wastes for the production of biomass and a rice crust potential applications.
- Bio-conversion of waste cellulose followed by the production of protein- rich microbial biomass of the product which decompose water.
- The use of microbial treatment to remove oil and metal chromium in the process of or remove the ammonia.

Technology vital health

It requires technical vital health in Iraq, in general great efforts to bridge the gap between developed and diagonals in the country to obtain the desired goal in health care.

- Establishment a center for bone marrow transplant in a hospital provides a number of specialists as well as equipment and materials.
- Start research projects in the field of gene manipulation in the tumors and other genetic diseases.
- Expansion of the pharmaceutical industry in the country and its development to meet local requirements as well as the modalities for the use of biotechnology.

Environment

The environment:

- Use of natural organisms (yeast, fungi and plants) to convert hazardous materials in the soil.
- Use of persistent micro- organism in the drainage systems for sewage and industrial sites as well as cleansing.
- Use of biotechnology to avoid contamination from the use of bioreactors for the products of hazardous materials.
- Forensic applications technology to forensic science.

Cooperation with international agencies

These include:

- Cooperation with the Islamic and international agencies that are required in the present time.
- Requiring trained scholars of Islamic and Arab countries that are directly involved in testing and the transfer of various biotechnologies.
- Development of short training courses for postgraduate students international organizations such as the UN agency of Educational and Cultural Organization (United Nations Educational Scientific and Cultural Education and UNESCO).
- Training of health personnel, especially doctors in the field of bone marrow transplantation and to assist in gene therapy, especially in cases of cancer of the blood and lymph gland tumor.

- Broadening the base of biotechnology such as the need to characterize the collection and preservation available mainly in the gene bank globally and provide information on all genes.
- Developing training programs for joint cooperation with various countries in the following fields in the area of biological control:
  - The exchange of biological information.
  - Mass production of the insect host.
  - Put out a vital crop for insects and the development of the joint projects.
  - The computerization of information and the development organization to link networks in different countries.
  - Training in the different features of the control of the House.

Foundation Center for Biotechnology in the light of Islamic treads provided and can be concluded that the status of biotechnology in the Islamic countries has become a necessity, especially those with deeper experience in biotechnology, which can greatly help in establishing such a center.

Proposal for the development of biotechnology in Iraq
- Applications in plants and agriculture
- Within the genetics

Proposal for the development of biotechnology in Iraq
Applications in plants and agriculture

The application of modern genetics, genetic engineering fundamentals of moving and developing improvements to allow farmers to produce new varieties and in various forms more quickly and at lower cost and in contrast to traditional methods that require time, effort. The followings are some applications of genetic engineering in plant improvement:

- Improving the quality of food, especially protein stored and other materials.
- The production of plants capable of nitrogen fixation and private non-legumes.
- Production of new crops resistant to:
  - Pesticides.

- Drought and therefore can be grown in deserts.
- Various pests and plant diseases
- Salinity and therefore can be grown in sea water.
- Frost in the country and in the cold winter nights.

- The transfer of genes between plants for the production of new types, especially the transfer of genes between species long descent through the development of new varieties.
- The production of plants with great ability in the operations photosynthesis.
- Cultivation of plant tissue in the hybrid types of long descent and even on the cellular mating in different ways.
- A plant products by micro- organisms through the transfer of plant genes and bring them into micro- organisms such as bacteria.
- The production of bio- pesticides "Production of Biocides".
- The production of bio fertilizers, "Production of Bio fertilizers".
- Tissue Culture "Tissue Culture Techniques".
  - Production of potato tubers and seedlings free of viral diseases.
  - Mass production of palm dates.
  - Production of the fruit crop.
  - Production of secondary materials.

Applications in animal

The progress that has been associated with new technologies such as artificial insemination and the use of female incubators and other techniques, especially genetic engineering to improve the offspring of animals through the production of useful new strains characterized by the following:

- Increase the number of resulting offspring.
- To contain higher amounts of red meat and with less fat.
- Its capacity for generating a great deal of milk with respect to types of livestock.
- Considerable weight and an increase in the number of eggs produced with respect to poultry.
- Specifications for the excellent wool sheep.

May contribute to genetic engineering in the area of changes in some animals in the following specifications:

- The possibility of transferring nuclei from one object to another.
- The use of livestock as a factory for the production of certain hormones and proteins that can be counted in milk and through:
  - Production of animals that carry transgenes known animals.
  - Either way, the other used to introduce genes into animals the use of viral vector known as the virus can be cultured bounced inside the desired gene is then inserted into the virus in animal cells to merge with the animal gene and begins to express itself within the animal.
  - Can take advantage of this method to introduce genes with essential medical value to humans.
  - The creation of a hormone in the cells of the individual in the dairy cattle only.
  - Production of some clotting factors (factor No. 9) responsible for the hemorrhagic diseases.
  - The production of many drugs such as interleukin-2 (to treat some types of cancer) in the rabbit.
  - Production of human growth hormone.
- Increased livestock productivity through:
  - Animal hormone injections.
  - The introduction of genes to increase the amount of hormones.
  - Increase the amount of wool.
  - Reduce the amount of fat.
- Building genetic maps of livestock, poultry and fish.
- Purification plant embryos in cattle.

Some applications in the industry

Of the most important applications in the industry to use objects allocated for Genetic Engineering to perform the following tasks:

- The production of antibiotics (the amount of larger, better quality, lower cost).
- The production of enzymes or different drugs that have medicinal value.

- Pollution control oil during the cracking of chemical compounds or remove pollution.
- Convert human waste to the food of a high content of protein for use in animal feed.
- Manufacturing and production of materials for energy such as ethanol, methanol and acetone.

In the treatment of oil- contaminated soil can develop a technology of high efficiency to help the oil sector to improve its products and expand, markets and increase prices and features of this technology as a clean environment and ensure the safety of its implementation contribute to economic return.

The project includes:

- To intensify and isolate bacterial strains that have the ability to remove sulfur.
- Increase the capacity of bacteria to remove sulfur and are counted at this stage, genetic engineering techniques, where is the transfer of specific genes carrying the desired qualities and planting, fast- growing bacteria plasmids.

Applications in humans

The study of genetic diseases is due to change in chromosome number or change in composition or changes at the molecular level of the gene, and is recognized as a gene that causes the disease gene through:

- The family.
- Use of modern techniques.
- Medical consultations: The family is a platform for diagnostic tests.
- Survey and genetic diagnosis of genetic diseases by:
  - Ultrasound.
  - Identify the presence of chromosomal changes or chemical weapons.
- New technologies

- Diagnosis and presence of specific genes in pre-reproductive individuals (which may be an individual carrier of the disease as a result of a new gene without the appearance of signs of illness it).
- The presence of the gene in the fetus before birth.

- Development of a sophisticated system for the diagnosis.
- Evolution of techniques of PCR.
- Develop the use of genetically modified animals in medical research.

Environment

- The use of vital signs "Bio-indicators" and controls the vital "Biomonitors" to detect the levels of pollution.
- Cracker diversity and the recycling of industrial waste and agricultural products.
- Development of resistant plants.

Other areas of research

- Technology enzymes
  - Enzymes for the diagnosis.
  - Clinical biochemistry.
  - Quality control of food.
  - Industrial processes.
  - Environmental control.

- Industrial enzymes
  - Design and analysis of bioreactors.
  - Waste water treatment.
  - The technology of the vaccine.

Transfer of biotechnology

The use of an information base for the diagnosis of "STR" (Short Tandem Repeats).

- Design prefixes "Primers Design".
- PCR amplification for the purpose of the STR.
- Analysis of the STR.

- PCR amplification of the mind "(SNPs) Single Nucleotide Polymorphism".
- A diagnosis of SNPs enzymes.
- A diagnosis of SNPs, respectively.
- PCR amplification.
- Genetic fingerprint "Finger Printing".

Chapter eight
Technological strategies

Entrance of the strategy

The strategy is a term originally used in the military sphere then the political sphere. It means the art of leadership, planning and coordination. Then it was used in the fields of education, management and others. Later, it became well-known in the field of higher education as the art that is administered by the (higher education), as planned, premeditated and directed purposefully. Moreover, the strategy is known by the "UNESCO" as comprehensive entrance for education that contained systems, curricula, programs, tools and its difficulties

The strategy of higher education is meant the set of and revenues in line with the targets set-ideas and principles that address this field in a comprehensive and integrated manner. Its major goals, objectives and targets derived from detailed specific principles, from which follow-up and evaluation methods are derived. Therefore, renewing and developing the strategy of Iraqi higher education includes:

- Develop philosophy and intellectual documents underlying them.
- Identify targets, goals, and objectives to be achieved.
- Content of the strategic plan.
- Creation of the means of implementation (physical and human) to achieve goals.
- Determination the time frames for completion.
- Evaluate the implementation and further development.

Higher education is a stronghold of human through, and the university is a scientific institution of the most important institutions of contemporary society associated with scientific development. Then, it requires the need to strengthen community support for the university materially and morally, whereas to hunt down the developments and assimilation of normalization according to specific priorities, including helping the Iraqi society to contribute to real competition in the global intellectual and scientific production. In order to achieve that, the Ministry must study the lessons of the past and take the reality into account. Naturally, the future will be the target of the planning, and the respected system of higher education in Iraq must stand to face the challenges and the rapid changes of circumstances. The slow progress of the higher education system is a proof of its ability of re-building, participation in the advancement of mankind. The aim of the creation of the strategy of higher education document is to design its outlines of restructuring and reconstruction.

The recent renaissance of the higher education system in Iraq is facing different trends and current tugging of various reforms. It is between this and that there must be clear strategy agreed upon to lead the pace towards achieving the goals of short-range or far. The document prepared is a vision for the near future, including the variables of Iraqi society undergoing planning that requires long-term stable solid vision, and without it no consideration can be extended beyond the possible.

To unite and clarify the concepts and terminology in this document, it is useful to indicate that the document used the term "educational institution": (Institution) to denote (Universities) and (Colleges), which grant bachelor degrees whether vocational or technical or professional institutes (Institutes) awards higher diploma or education, and those associated with universities or other related ministries. The use of the terms (institutes) and (institutions) of higher education will mean those associated exclusively with the Ministry. Different meanings of terms: "Globalization" and (Internationalization) will be a distinction between their implications as appropriate.

The document is summarizing some of the most important center for social and political conditions that have affected, and continue to affect, the march of Higher Education and Scientific Research of Iraq. It has then submitted the reflecting of the most important principles that could be based upon a vision for the future. Thus, making intellectual and philosophical priorities is set. Then, comes the drawing of a strategic plan that targets ten years without risking the visions fore the future of its features.

Strategic vision and future renewal of higher education in Iraq

This document is not designed to interfere into the circumstances of events and how the higher education system got up. It is designed to demonstrate and then outline a strategy to restructure and reconstruction.

This document is a novel for the near future, including the variables of Iraqi society being what it is painful because it is recognized that planning requires long-term stable vision that provides a solid stronghold vision without which we can not look beyond the possible.

Iraq has suffered over decades of war and siege, resulting in loss of life and material damage, beyond the individual to society at large and even land, but the biggest damage is the missed opportunities for progress, loss and the emergence of several phenomena, notable missing the learning opportunities for children, high child mortality rates, low levels of growth and missed opportunities for construction.

The document ascertain the collapse (almost universal) in the educational system, such as the continuing departure of the academic, the country's asylum from death threats, which caused great destruction to the infrastructures. However, the educational system is currently undergoing a major phase in the rebuilding, due to the efforts of the Ministry, constant changes in its institutions, increasing the effectiveness of what regards the development of curricula, assessment and quality of scientific research and technical education, academic freedom, decentralization of educational, welfare of teachers, students and the permanency of funding. The challenges facing Iraq can not be addressed in part, but must coexist with strong influence in Iraqi society perceptively. These challenges are: economic reform, demographic change, communication with the outside world, globalization, and rapid increase of knowledge. In the face of these challenges, reorganization of higher education activity in the light of basic vision of the future and the general trends of the strategic plan (in a real short-term goals), which is targeted by this document, are to prioritize the order, in order to invest the available sources based on mental fact effectively and influentially.

New legal and academic vision of the strategy

The new vision of law must be founded on the previous law, with the preservation of the original features, which should remain as part of the educational and cultural heritage, taking into consideration the rapid chances required by the new society with harmonious development in higher education. It should be more responsive to the problems faced by the needs of Iraqi society, and closer to the problems of diverse geographical regions. On that basis, the universities and scientific institutes should be regarded as scientific institutions by Iraqi laparoscopic and as follows:

• Made it to be a contribution to enrich the social and economic change and the promise of sustainable human development.

• Made its contribution to organize the advanced society and closely related to the search for fighting poverty and protecting the environment, improve public health and nutrition, strengthen the principles of civil society and is keen to develop the levels and other types of education.

• Made it responsive to changes in the labor market and civic culture and therefore, attempting to develop the potential of civil society groups and predisposing own quest for the development of individual competencies .

The new strategic vision should look to the universities and institutes of higher education as pioneers for the development of the trends of promising players towards the labor market and lead to the emergence of new opportunities for future development. Resulting in greater attention to major changes in the market, adoption of curricula and education systems in line with changing circumstances based on principles of credibility of the quality and management of institutions of higher learning. New vision embodied in this regard must ensure the legal autonomy of institutions of higher education both formal and informal for the opportunity to be creative in society. The practice of giving autonomy to educational institutions would hold those institutions greater responsibility in academic work and its contents as well as ethical issues of exchange and expenditure.

The document was presented on the search for outlets for dealing with these challenges, which has become inevitable, the deal with the administrative weaknesses is an important issue in this regard and it is in the interest of higher education (official and private), that takes into account the awareness and recognition issues, including educational institutions as part of building a respected management system. Among the elements that should pay attention to this vision is the renewal of teaching methods and contents.

To reach those goals according to the future renewal of higher education, the new vision should draw the image of higher education system in Iraq as shown in the following:

- Training system of high-quality active and influential in the framework of civic and professional activities, including the developments varied specialist.
- The system allows easy reach to the sources of science and to interact with its programs so as to ensure equality and social welfare.
- Education system is based on the awareness and knowledge alone. It encourages commitment to keep and focus the trends of knowledge in the minds of future graduates as well as a sense of responsibility for training in the service of social development.
- System in which to discuss local, national and global issues by the spirit of criticism and encouraging citizen participation in issues of culture and civilization.
- System easy for the government and other educational institutions to obtain access to the information required by the documented decision-making process.
- Social environment engaged in research innovation, knowledge, creation and evolution of scientific and technological development.
- Social environment encourages coordination with the industrial sector and service in order to develop the national economy.
- Social environment where its members are interested in democracy, human rights, social justice and building the culture of Islam.

Platform of the Strategic Plan for Higher Education

The strategy of higher education in Iraq depends on:
- Higher education is a comprehensive system.
- Higher education system has several links to economic, social, political and cultural right systems.
- Iraq has glorious heritage with interactive role in human life.

This plan is composed of the followings:

- Institutional building of higher education. The following questions have to be answered:
  - Who will decide the infrastructure of higher education?
  - What are the qualifications of higher education?
  - What kind of institutions that will be joined under the banner of higher education?
  - What are the criteria that will govern the construction and conversion of the various institutions of higher education? What are the criteria that will govern the organization of educational institutions and fields of specialization in the Iraqi higher education?
- The administrative system. The following questions have to be answered:
  - What is the balance between the government authority and the institutions of higher education?
  - What are the principles and the environment? Which will represent the administration at the national level?
  - How will be the design of the form of management and leadership at the level of educational institution?
  - How will be the management structure and research centers?
- Organization and attraction of scientific cadres: the answers of the followings are required:
  - Who will be responsible for creation, transfer, management of scientific centers and how the planning is done to attract academic staff?
  - How the process of improving conditions for staff career will be managed?
  - How will the evaluation and stimulation of teaching staff, promotion of academics be carried out?
  - How will the development of qualifications and the adequacy of Iraqi cadres be accomplished?
- Admission and care for students: the questions in this regard are the following:
  - What socio-economic variables that will have impact on higher education?

- How would the ideal choice and standards for students' admission be selected?

- What are the services to be provided to students?

- What are the activities needed to enrich the academic performance of students?

• The qualitative aspect of training: the following questions are raised:

- How will the relationship between teaching and scientific research be?

- How will the future of the curriculum be?

- What would the system of college hours and evaluations be?

- What are the qualifications required to obtain certificates, and what are the contents of training programs in the Iraqi higher education?

- How will the system of quality control and recognition of academic be built? What are the criteria adopted in granting leave to foreign donors?

• Physical infrastructure: what are the needs of data, devices, dormitories, sports supplies and nominal programs required for the buildings?

• The management and funding of higher education:

- Who will be funding the Iraqi higher educations?

- Could the current situation be reformed?

- How the funding allocations of educational institutions could be improved?

This document shows details of the events and projects which are included in the plan. This requires a range of measures on several things including the reconstruction of buildings, attraction and development of scientific and administrative staffing, acquisition of necessary equipment for laboratories, equipping libraries inputs, teaching techniques, modern processing techniques, training to deepen the learning of students and others.

This document also contains the main lines of strategy of national higher education, such as the application of university service law and the reform of graduate and under-graduate studies. Then the development renovation and modernization project of the Institute of Higher Education particularly in areas that suffer from delayed higher education services. As well as the establishment of a center for the measurement and evaluation and training center for students, the Supreme Iraqi Foundation for Scientific Research and an Iraqi academic for recognition.

In the area of expansion of students' admission to meet the needs of the community of skills required, takes into account:

- The involvement of female in higher education, especially in disadvantaged groups and regions with economic needs.
- Sustaining and supporting programs for the development of academic staff, technicians and administrators qualitatively and quantitatively, specifically in:

  - Updating the skills of teachers' in-service training courses.
  - Using missions, twinning, and inviting Iraqi specialists abroad to work in higher education through programs of the United Nations Development Program (UNDP).
  - Transfer of knowledge through the Iraqis (TOKEN).
  - Coordination and cooperation with higher education systems in the Arab countries.
  - Establish body for university grants and fellowships.
  - Improve living conditions for workers and students.
  - Establish a communication network for universities and institutes of higher education.
  - Establish Iraqi information management systems to improve the management of Higher Education.

- Balance between the level of output and employment market
- Continued dialogue between departments at different levels.
- Flexibility and knowing of the characteristics of the external environment.
- Taking into account all aspects of the internal environment of the institution of the university.
- The interaction between planning and execution.

The re-identifying steps of strategic planning are important to characteristics of university environment, and analysis of internal and external academic institution. But, there is a number of defects that affect the strategy for higher education, including lack of awareness among future rights that leads to problematic practice, weakness of the strategic work of faith in higher education institutions, and weak organizational support for government action.

Higher education strategies

The main issues needs and priorities that emerged from the UNESCO Roundtable were the following:

- The wide spread destruction of the infrastructure of the higher education system.
- The unstable and dangerous environment for normal academic activity.
- The quality of higher education has been steadily deteriorating since the imposition of authorarition rule in Iraq.
- There is a need to equip more than 2000 scientific laboratories and for 30000 computers. Libraries are in a poor condition and are in urgent need for restocking with new books and journals in both Arabic and English; journals in electronic format are required.
- The student population has been rapidly increasing due to both a high birthrate and an admission policy that allows all students who have completed secondary school to enter higher education.

To reconstruct the higher education, it is required to: put a spanning plan (5-10 years), maximize higher education graduation, direct students opportunities in learning towards the public needs, develop a team to inspire students and teachers to work in teams, adopt curricula upon bases of appropriate outcomes and capabilities, develop teaching materials, class size should be designed to suit the student learning process, and put enrollment policies that aim at recruiting only enthusiastic students.

Consideration needs to be given to the relationships between higher education and industry and commerce. The need to update the information in higher education, information and learning centers should be established. A creative policy is needed for the provision of learning resources.

The Strategic Action Plan for Higher Education in Iraq should consist of two major parts.

Part One:-

A. Institutional Structure of Higher Education

- Who should decide on the basic structure of higher education?
- What should be termed "higher education"?
- What types of institutions should higher education encompass?
- What criteria should guide the establishment and allocation of tasks between different institutions of higher education?
    a. Consolidation
    b. Specialization
    c. Integration
- What criteria should guide the organization of institutions and fields of study in Iraqi higher education?

B. System of Governance

- What should be the balance of power between the national authority and the institutions?
- What should be the principles and structure for governance at the national level?
- How should organization and leadership be set up at the institutional level?
- What should be the governance structure for research?

C. Students' Admission, Retention and Welfare

- What are the socio-economic developments that will have an impact on higher education?
- How can access of girls and disadvantaged groups in higher education be encouraged and increased?
- What should be the appropriate criteria and the system for the selection of students?
- What types of student services will need to be developed
- What types of activities will enhance student achievements.

D. Organization, Recruitment and Training of Academic Staff

- Who should be responsible for the creation, allocation and management of posts and positions?
- How should one plan for the recruitment of academic staff?
- How should academic personnel be recruited?
- How careers and employment conditions should be managed?
- How should academic staff be evaluated/appraised and motivated?
- How could the qualifications and competencies of Iraqi staff be improved?
- How should the skills of the academic, administrative staff and technicians be refreshed and next generation is trained?

E. Quality of Teaching and Learning

- What should be the relationship between teaching and research?
- What should be future structure of study programmes?
- What should be the system of credits and evaluation?
- What should be requirements for certification?
- How should the contents of training programmes in Iraqi higher education be revised?
- How should a sustainable system for quality control and accreditation be created?
- What should be the criteria for the licensing of foreign providers?

F. Physical Resources and Structures

- What is needed for:
a. Physical plant
b. Equipment
c. Students hostels (dormitories)
b. Recreation facilities.
- What building programmes need to be envisaged?

- How can a system for management, procurement and maintenance be set up?

G. **Management and Finance of Higher Education**

- Who will finance Higher Education
- How can current status of financing contribute to achieve reform?
- How can allocation of funds to institutions be improved?

Part Two:

The Strategic Action Plan should detail the many projects activities required to realize the plan. Both parts must be evaluated and approved jointly by the Ministry of Higher Education and Scientific Research (MOHESR); Ministry of Economic Planning (MOEP); and Ministry of Finance (MOF).

In addition to shaping a standardized system, immediate action needs to be taken to implement the minimum conditions allowing the higher education sub-system to operate at an acceptable level. This entails suggestions related to the refurbishment of buildings, recruitment and upgrading of academic and administrative staff, provision (purchase) of essential equipment and supplies for laboratories, libraries and teaching-learning materials, training materials to enhance immediately student learning. The main line of the National Strategy for Higher Education can be outlined as follows: (For further details, readers are advised to refer to plan of actions tables.)

❖ Implement Universities Service Law.

❖ Revitalization of under-graduate and post-graduate colleges.

❖ Establishment of new under-graduate and post-graduate.

❖ Renovating, up-grading and re-structuring existing institutions of higher education with special emphasis on the regional universities, particularly in those geographical areas where provision for higher education is relatively scarce .

❖ Setting-up an, Evaluation and Measurement Centre; a Post-Graduate Training Centre; an Iraqi Research Foundation; and an Iraqi Accreditation Agency.

❖ Expanding student enrolment to meet social demand, economic needs for skills and equity in respect to (a) female participation; (b) disadvantaged groups; and (c) economically deprived and under-developed regions.

❖ Setting-up/strengthening staff development programmes – academic, managerial, and technical – in quantity and quality, in particular, upgrading academic staff members through on-campus training but also through provision of a scholarship programme and twining universities.

❖ Inviting the Iraqi Diaspora for Higher Education development through the UNDP project TOKTEN (Faculty Support and Curriculum) and recruiting returnee Iraqi academic staff.

❖ Setting-up of the University Grants Commission.

❖ Providing improved living and working conditions for both students and staff.

❖ Installing Networks of Higher Education Institutions.

❖ Establishment of the Higher Education Management Information System.

Building strategy principle for higher education in Iraq

The building of a strategy for higher education in Iraq requires the clarity of strategic principles, ways of implementing such as: strategies concepts, sources, and identification of the challenges of domestic and external nature. The implementation requires clarity stages with the standards of priorities.

The major tasks to develop a strategy for higher education is the clarity of its principles and objectives, detailed procedural, derived from the philosophy of education, depends on the clear social philosophy, based on the heritage of the nation and the living past and current realities and global challenges. Accordingly, the success of the strategy depends on the availability of key principles that could be adopted as initial principles of the social philosophy of education. These principles can be summarized as the following indicators:

Humanitarian indicators
- Development of the personality of the learner
- The rights of original human learner
- Reliance on self-help
- Indicators of faith
- Consolidation of belief in God
- Establishment of laws on personal rights
- Human values
- Humanitarian brotherhood
- National indicators
- Patriotic spirit
- Adherence to the national territory

Ways to implement the strategy

Several actions are required in implementing the strategy, including:
- Reopening the debate and dialogue in universities and scientific research centers and workers in the fields of comprehensive development and the dialogue include:
  - Intellectual aspects of the strategy

- Practical aspects of the strategy
- Development of the existing institutions.
- Adoption of models for the implementation of particular priorities.
- Difference in priorities of what has been found in other states (financial problems, providing manpower).
- Limitation of financial, human and material resources.
- Linking plans and educational projects, by plans and objectives of comprehensive development projects.

Higher education in Iraq is facing domestic and external challenges involved together with the world, characterized by special challenges, including:
- Scientific and technological revolution.
- The production of knowledge.
- Economic changes.
- Political changes.
- Cultural changes.
- Social changes.
- Globalization.

Stages of the implementation of the strategy

There are key stages of implementing the strategy that vary from country to country and precedes these stages a number of required actions, followed by evaluation and review all of these elements can be summarized as:
- Phase of creating requirements.
- Stage of procedural setting goals.
- Stage of development of operational plans.
- Phase of the implementation of procedures and processes.
- Phase of detailed implementation plan.
- Description and analysis of output. Evaluation and follow-up phase.

To develop a specific time for each stage and to implement the strategy are essential for each country. The officials may not be able to meet the requirements of this point in time they may also lead to delays in implementation or planning or accelerating in some aspects that require precision. However, we should identify indicators of time to help to find the road when implementing the strategy.

It is necessary to identify priorities, which prefer to begin with implementing the beginnings or the fundamental premises of the strategy. The priorities can be summarized and identified as follows:

- Difficulty and suffering of priority.
- Participation of priority in all sectors of society.
- Linking priority to provide manpower.

According to these standards, priorities can be identified, for example, as diversification and development of technical education because of its importance in providing the labor market by frameworks.

Strategic components of higher education in Iraq

These components of strategy for higher education in Iraq include topics of interest problems based on problems mentioned above that require solutions, and from these:

- Admission
- Curricula
- Development of human resources
- Graduate studies
- University administration
- Quality assurance
- Financing of universities
- Legislation

The current historic juncture in the life of Iraq is witnessing a new society building of course. What has happened in April 2003, introduced a fateful turning point in Iraq in its history necessitated building or rebuilding of all joints of life.

In the midst of wrestling political, intellectual and cultural philosophies, is opened up in larger parts in Iraq. The planner is accustomed to find difficulties in choosing or to reaching a solid philosophical features that combine all these conflicting trends (and sometimes contradictory), and then charting the vision for the future. The strategic decisions are turning to be justified logically; agreed to constants; find only a single supplier and reliable coverage for most of the conflicting reliable trends, namely the new Iraqi constitution.

Cornerstones of the constitutional strategy

The constitution is relatively stable with high acceptance. Therefore, this document (constitution) a strategy of higher education will be based upon for years to come because it is characterized by stability, as well as clear legislative texts. The document can be inferred to many indicators, as explained in the following:

Although the preamble of the constitution is not a part of its legal rule, it can provide for the tracker a reading of a comprehensive social development which resulted from the constitution. They have been at the forefront among other signals including the following:

- Iraq is the home of the apostles and prophets and the cradle of civilizationsThe Iraqis are write-makers and advocates of agriculture and manufactures of numbering.
- Iraqis testified many repressions.
- Iraq consists of several components.
- The new Iraq (Iraq of the future) is not sectarian or racist or co-notational or regional or discriminating.

These fundamentals of the scheme explain the ground of psychological, social and cultural development of the culture of the current stage. In the hopes of higher education in this perspective, we have to give the following principles and concepts shared by the majority of higher education systems counterpart:

• The principle of intellectual freedom: the suffering of the Iraqi thinker of authoritarianism and individual and intellectual siege that create a strong tendency towards academic studies to be enjoyed by all educational and research institutions.

• The principle of independence of scientific research which distinguishes scientific research from any activity or product of human thinking is the principle of objectivity and impartiality.

• The status of higher education: supposed to be at the forefront of higher education community in reading and future requirements of Orientalism and preaching the following scientific knowledge. It must take its place in high-interest country, and has a whole-care necessary to achieve this.

The responsibility of higher education: higher education bears responsibility for humanity and patriotism towards the country and contributes towards the humanitarian needs, and supplies the needs of Iraqi society manpower, at the same time contributes to supply the human knowledge development.

Relationship with higher education of the world: Iraq contributes to accumulation of intellectual and cognitive taking and giving. It shows the concepts of modernization and global dissemination of culture and human knowledge.

Aspects of strategic plan

The strategic plan for higher education for the next decade covers seven specific areas that represent the most important activities of the Ministry of Higher Education and Scientific Research. They are the elements of the educational process, and they are centered on human resources, material resources, technologies, management system, finance, and quality. The inclusion of these themes represents the questions and then the answers which represent the strategic objectives, events and activities planned for their achievement, and as follows

- Organizational structure and institutional building for higher education

-Determination of validity of the body that decides the infrastructure for higher education and its administrative configuration.

-Identification of higher education.

-Institutes that will join under the higher education.

-Standards that will govern the association and the financing of various institutes of higher education.

-Standards that will govern the relationship of the organization upon the Iraqi higher education.

-The required balance between government authority and institutions of higher education.

- Student affairs

-Opportunities for females and the disadvantaged enrollment in higher education.

-Perfect system for selecting students and standards.

-Services to be provided to students.

-Activities needed to enrich the student's academic performance.

- Human resources

-Planning to attract academic staff.

-Management of the process of improving conditions for career staff.

-Evaluation, promotion and stimulating of academic staff.

-Developing qualifications and competence of Iraqi cadres.

-Developing the skills of personnel (teachers, technicians and administrators).

- School system

-Relationship between teaching and scientific research.

-The future curricula.

-System of school hours and evaluation.

-Qualifications required for obtaining certificates.

-Academic recognition.

- Physical infrastructure

- Requirements for buildings, appliances, dormitories equipment.
- Programs required for designing buildings.
- Building an effective system for maintenance.
- Funding and financial management
- Iraqi higher education funding.
- Reforming of the current financial situation.
- Improving the funding allocations of scientific institutions.
- Dimensions of investment and productivity of educational institutions.
- Maintaining the quality and assurance
- The process of evaluating students.
- Criteria for evaluating the performance and quality measurement.

strategic plan for higher education in Iraq

The goal of activities and events is to achieve the quantitative targets within time ceiling priority.

- Reform the system of higher education
- Completion and adoption of the laws of higher education.
- Symposia and conferences to demonstrate the prospects for development and reform in 2006.
- Agreement on the structure and institutions of higher education and reform
- Studying the proximate merging of institutions geographically.
- Studying linking educational institutions with technical university nearest suitable.
- Adoption of colleges and institutes.
- Studying Iraq's needs for universities developed in 2006.
- Building a system of distance education
- Determination of appropriate required disciplinary system.
- Preparation of the criteria for evaluating the distance education.
- Qualified staffing members.
- Creating appropriate educational items
- Extensive media.

- Development of a program for distance education at university in 2006 as a pilot.

- Teaching system of foreign universities
- Overtures and negotiations with donors and specialized parties.
- Twinning of Iraqi universities with their foreign counterparts.
- Development of realistic programs to teach foreign languages.
- Various mechanisms for enforcing the law on higher education and university
- Management of organs development of universities and configurations.
- Organizing a national conference of leaders of higher education.
- Specializing of the founding universities
- Studying the needs of the geographical areas of specialization.
- Establishment of universities in different areas as social and economic needs 2006.

Document of strategy of higher education in Iraq

The document of the strategy is aimed to upgrade the higher education through long-term future plan of (10-25) years, with distinguished scientific ambitions vision, clear message with impressive values and standards to evaluate a variety of higher education systems, and patterns and methods of financing. This holistic strategy included several issues: the admission and coordination with the labor market, finance, infrastructure, management, postgraduate studies and scientific research. Accordingly, it is needed to create a scientific base to absorb modern science, technology and work to develop it. In addition, the continuation of this effort is carried out by the help of distinguished researchers. The creation of scientific structure, researchers, research centers and laboratories, will enable Iraq to contribute to modern scientific progress and phasing out liability to scientific and technology capabilities and experience in introducing modern technology to local markets. In this context, it is not possible for Iraq to benefit from the capabilities of its migrant scientists unless it can create an environment capable of internal scientific temptation to return to Iraq; or at least hiring sciences, knowledge and scientific research.

The creation of an Iraqi scientific community believes in the importance of science, and in the decentralization of planning that requires long-term patterns of thinking and multiple directions for the development of a society that is able to understand the central importance of science and scientific research, the reform of higher education system in Iraq, the development of teaching methods and curricula by encouraging scientific research and creating team spirit through collective research. The need to provide incentive for TAPGUN scientifically as well as starting in scientific research means adopting policies of the value of science and scientists in the community and provide a decent life for researchers and scientists.

Strategic planning of the future of traditional higher education in Iraq includes the following specifications:

- Control of management of available resources within the university.
- Control the factors of external environment for academic institution.

Higher education Policy and strategies

The main issues needs and priorities that emerged from the UNESCO Roundtable were the following:

- The wide spread destruction of the infrastructure of the higher education system.
- The unstable and dangerous environment for normal academic activity.
- The quality of higher education has been steadily deteriorating since the imposition of authorarition rule in Iraq.
- There is a need to equip more than 2000 scientific laboratories and for 30000 computers, libraries are in a poor condition and are in urgent need for restocking with new books and journals in both Arabic and English; journals in electronic format are required.
- The student population has been rapidly increasing due both to a high birthrate and an admission policy that allows all students who have completed secondary school to enter higher education.

To reconstruct the higher education, it is required to plan spanning (5-10 years), maximization higher education graduation, learning students opportunity according to the needs, development a team to inspire student, teachers to work in team, curricula should be based on appropriate outcomes and capabilities, development of teaching materials, class size should be designed to suit the student learning process, and the enrollment polices need to aim at recruiting enthusiastic students.

Consideration needs to be given to the relationships between higher education and industry and commerce. The need to up – to date information in higher education, information and learning centers should be established. A creative policy is needed for the provision of learning resources.

The Strategic Action Plan should detail the many projects activities required to realize the plan. Both parts must be evaluated and approved jointly by the Ministry of Higher Education and Scientific Research (MOHESR); Ministry of Economic Planning (MOEP); and Ministry of Finance (MOF).

In addition to shaping a standardized system, immediate action needs to be taken to implement the minimum conditions allowing the higher education sub-system to operate at an acceptable level. This entails suggestions related to the refurbishment of buildings, recruitment and upgrading of academic and administrative staff, provision (purchase) of essential equipment and supplies for laboratories, libraries and teaching-learning materials, training materials to enhance immediately student learning. The main line of the National Strategy for Higher Education can be outlined as follows: (For further details, readers are advised to refer to plan of actions tables).

- ❖ Implement Universities Service Law.
- ❖ Revitalization of under-graduate and post-graduate colleges.
- ❖ Establishment of new under-graduate and post-graduate.
- ❖ Renovating, up-grading and re-structuring existing institutions of higher education with special emphasis on the regional universities,

particularly in those geographical areas where provision for higher education is relatively scarce.

❖ Setting-up an, Evaluation and Measurement Centre; a Post-Graduate Training Centre; an Iraqi Research Foundation; and an Iraqi Accreditation Agency.

❖ Expanding student enrolment to meet social demand, economic needs for skills and equity in respect to (a) female participation; (b) disadvantaged groups; and (c) economically deprived and under-developed regions.

❖ Setting-up/strengthening staff development programmes - academic, managerial, and technical in quantity and quality, in particular, upgrading academic staff members through on-campus training but also through provision of a scholarship programme and twining universities.

❖ Inviting the Iraqi Diaspora for Higher Education development through the UNDP project TOKTEN (Faculty Support and Curriculum) and recruiting returnee Iraqi academic staff.

❖ Setting-up of the University Grants Commission.

❖ Providing improved living and working conditions for both students and staff.

❖ Installing Networks of Higher Education Institutions.

❖ Establishment of the Higher Education Management Information System.

Edcational strategy adopted

The implementation of some of the strategies that led to alleviate the pressures and limitations suffered by the education system. However, the challenges and the workload were enormous and the resources specific to that. There was an urgent need to identify priorities that had to be influential actors strategies, to improve the quality and efficiency of education for all and with same resources available. The appointment of priorities properly and adopt reasonable methods that limit the cost were very necessary and also to learn from international experience, particularly from countries with similar resources. The most prominent foundations, strategies and policies for the education sector that have been adopted included the following:

• The pursuiting of quality education is reflected on the special education system and the inputs and outputs of all of its operations, allowing the opportunity to contribute to the desired change in society.

- Providing educational opportunities for all and eliminating leakage of learners from all levels of education and spread of lifelong education.
- Eliminating disparities in the enrollment of teachers for boys and girls, rural and urban areas, between various ethnic backgrounds and different economic conditions.
- Improving the quality of education in order to have better respond to the needs of the labor market, sustainable development catching up with countries with high-performance in the field of education, improving capacity and, efficiency of the positions of the teaching staff and faculty.
- Adoption of the independence of education, its separation from politics, respect for freedom of thought and expression and to promote tolerance and social cohesion.
- Promotion community involvement in planning, evaluating the educational system, strengthening coordination with higher education and other sectors and the development of private sector participation.
- Adoption of the scientific method in the educational system planning, implementation and assessment on the basis of an educational database structure.
- Reviewing and development of legislation dealing with the various educational components of the educational system, modernization of its activities in line with the requirements of educational development and keep pace with new developments in education.
- Strengthening the trend towards decentralization in education and the adoption of the school as a first essential to bring about educational and desired social development.
- Facilitating the ways to use information technology in the areas of education, administration, oriented towards building an integrated information system and linking the educational status of the ministry directorates of education in the field.
- Raising the profile of the teacher scientifically, socially, and economically, due its role in building human and society and the pursuiting of continuous professional growth.
- Development of curricula and textbooks to suit the changing needs of the individual and society and follow-up to the introduction of educational concepts, including the contemporary education.

- Development of educational techniques to cope with technological developments and uses it as an integral part of the curriculum.
- Development of school buildings as a kind, preparing and dispensing the school buildings, with the double- time and providing a supportive environment for the educational process.

To apply these strategies, many books have been written and forwarded to the printing such as Arabic literature and civic education, social and language development and English, with the revision and amendments to the other books, such as books and history of Islamic education. National Commission for curriculum reform and development have been formulated with a number of educational and scientific figures in the country and undertaken the task of the gears in the quality of curricula and textbooks, particularly those related to the language, mathematics. The provision of educational technology in teaching has been associated with the ministry during the 2004 - 2005 printed more than 80 million copies of books for different stages through the World Bank project and the potential for ministry self-review and amend a number of dimensions and concepts.

Strategy for Education in Iraq

- Providing an initial evaluation of the situation of education in Iraq
  - Developing of human resources and education
  - Looking at the impact of education
  - Funding of education
- The contents of the education strategy and education in Iraq
  - Principles of Strategy for Education in Iraq
  - Ways and stages of implementation of the Strategy for Education
  - Priority

- The general framework of the strategy of education in Iraq
  - Legal vision of the Strategic Plan
  - Intellectual pillars of the Strategic Plan
  - The constitutional underpinnings of the Strategic Plan
- Towards a strategy of Iraqi education
  - Requirements for future strategy
  - Axes of the Strategic Plan

Ways and stages of implementation of the strategy for Education

Requires several procedures at the implementation of the strategy is called the ways of implementing the strategy, including:

- Openness discussion and dialogue in educational institutions and workers in the fields of overall development and includes the interview:
  - The intellectual aspects of the strategy.
  - Practical aspects of the strategy.
- Development or improvement of existing institutions.
- The adoption of models to guide the implementation of the priorities for particular.
- The different priorities of what is found in other countries (financial problems, the provision of the workforce).
- Inventory of available financial, human and material resources.
- Linking plans and educational projects, plans and objectives of comprehensive development projects.

Phases are essential to the implementation of the strategy vary from one country to another before these stages, a number of required procedures followed by the process of assessment, review, and can be summarized as follows:

- The stage of creating the requirements.
- The stage of defining the process goals.
- The stage of formulating operational plans.
- The implementation phase of the proceedings and processes.
- Stage of the implementation of the detailed plan.
- A description and analysis of the output.
- An evaluation and follow- up.

That the status of certain periods of time to implement the strategy is necessary for each country, could face officials able to meet the requirements of this point in time, may also lead to delays in implementation, planning, or to speed up the areas that would require accurate however, a need to identify indicators of time to find a way of implementation of the strategy.

Priority

Need to set initial priorities preferably put into effect and considered the beginnings or the basic premises of the implementation of the strategy and priorities can be summarized and defined as follows:

- The difficulty of priority and suffering from them.
- The priority and presence in earnest.
- Participation in all priority sectors of society.
- Linking high importance to provide manpower.

The general framework of the strategy of education in Iraq

Iraq after decades of war, the embargo, the resulting losses in lives and material damage exceeded the individual to include the community at large and even the ground, but the biggest impact in the loss of opportunities for progress, loss and the emergence of the phenomena of many who have missed the most important learning opportunities, low levels of growth and loss opportunities for construction.

The strategy proposed that the state had witnessed an almost total collapse in the education system which has not been seen previously, as a continuation to leave academics to the country of asylum or otherwise, death threats and destruction of significant progress, including in infrastructure, however, the educational system is currently undergoing major reconstruction and refers to the efforts of the Ministry about the ongoing changes in its affiliated institutions and increasing their efficiency as well as on developing the curriculum, evaluation and the quality of scientific research and technical education and academic freedom, decentralized education and welfare of teachers, students and the sustainability of funding and spread globally.

The challenges facing Iraq can not be dealt with in part but must co-exist with the major players in Iraqi society, perspective in the near future as an economic reform and demographic change and communication with the outside world, globalization and the rapid rise of knowledge. The face of these challenges is necessary to regulate the working environment in the basic education activity, the light of the future vision and strategic parameters of the plan (in the perspective of short- term goals of the fact), a target of this document and the order of priorities for the investment of available sources on the ground indicating the mental and in an active and influential.

Legal vision of the strategic plan

Any legalistic vision must be based on a legal basis to maintain a balanced take into account the features of the original that should be kept as part of the educational and cultural heritage. With the introduction of changes required by the new society with harmonious development in education and make it more responsive to the general problems facing Iraqi society and economic needs, cultural and closer to the problems detailed geographical areas and diverse population based on this should be seen to universities and scientific institutes Iraqi arthroscopic the following:

- It is scientific institutions contribute to its tender, offer to enrich the socio- economic change and promise of sustainable human development.

- It provides scientific institutions bid for contributing to the organization of a developed society and is closely related to seek to fight poverty, protect environment, improve public health and nutrition and strengthening the principles of civil society and is keen to develop standards and other types of education.

- It is scientific institutions respond to changes in the labor market and civic culture, and then it seeks to develop the potential of civil society and qualified for special groups for the development of individual competency.

The vision of the new strategy have to look at universities and institutes of higher education pioneers the development of promising trends towards active labor market and lead to the emergence of new opportunities for future development which results in greater attention to major changes in the market, the adoption of curricula and education systems in line with the changing conditions based on the principles of quality and management of education institutions involved as much as on the performance of its role in the creative community. As well as the practice of giving autonomy to educational institutions that applied formulas with legal fact, and certainly on individual freedom and independence of the university, which would pass through these institutions assume greater responsibility in academic work and contents of ethical issues, as well as exchange and spending.

The strategic view on the search for outlets to deal with these challenges and which has become inevitable, is no secret that to deal with administrative weakness is important in this regard and therefore in the interest of education (formal and private) that takes into account the issues of assessment and awareness, including issues of recognition of educational institutions as an important part of building management systems. Among the elements that should be given special attention in this vision is to renew the teaching methods and contents.

In order to reach those goals according to the vision of the future and for the renewal of education, a new vision that draws upon the education system in Iraq in a show that:

- A system of high- quality training enables students to deal effectively and influential within the framework of civic and professional activities, including developments in the various specialized.

- A system that allows easily to reach sources of science and to interact with its programs so as to ensure unlawfulness and social welfare.

- Education system is based on the quality and knowledge and attitudes alone and encourages knowledge and commit to keeping the focus in the minds of future graduates, as well as a sense of responsibility for the development of training in the service of social development.

Intellectual underpinnings of the Strategic Plan

This historical stage of Iraq witnessing the building of a new society and, of course, what happened in April 2003 (and also suffered from the events) that had been plunged Iraq into a decisive turning point in its history, summoned the building (or rebuilding) all joints of life.

In the midst of political philosophies, intellectual and cultural conflict and jam between the cultures, which was opened by the Iraq (or opened it), the chart find himself in great difficulty in the selection of the parameters, or failed to collect a solid philosophical these trends are often conflicting (and sometimes contradictory) to lean upon in making a vision insight into the future. Taking necessary for the strategic decisions that justify logically located where the constants agreed, but did not find a single resource can be used as a basis for its comprehensiveness, and that most of the conflicting trends reliably, but a new Iraqi constitution.

On the educational system directly in the new Iraqi constitution, which can be considered as an initial basis of the strategy is limited m particular in article 34:

- Education is a main factor for the progress of society, a right guaranteed by the state, which is compulsory at the primary level and the state guarantees fighting illiteracy.

- Free education is a right for all Iraqis in all its stages.

- To encourage the State scientific research for peaceful purposes that serve humanity and it promotes excelling, creativity and innovation and the various manifestations of excellence.

Draft strategy for the Iraqi education

The idea of the educational strategy in Iraq is attributed to:

- Directive of the Council of Ministers the need to develop a strategy for educational development.
- The formation of several action teams to study the reality of education and the challenges it faces.
- A draft strategic thrusts of education development.
- Preparation of the strategy document consisting of several axes.
- Presentation of the project on legislative bodies.
- The decision of the Council of Ministers that a draft strategy education.
- Constitute a task force and writing documents on axes and conduct field studies and follow up the implementation of this idea after it became a reality.

The importance of strategic planning for education

Strategic planning is long- range planning suggests any internal and external variables and identifies the target channels, the adoption of the best methods of competition and are updated as renewed factors external and internal.

Strategic planning aims to analyze the ingredients according to the approved channels planning process by the adoption of the scientific basis requirements and models are supported with forecasts and prepared, and implemented and evaluated according to a systematic strategy is clear.

The integrated elements of the strategic plan and based on scientific methods can be mentioned:
- Realism
- Reliability Flexibility
- Inclusiveness
- Democracy
- Efficiency
- Integration
- Continuity

Realism is appropriate to the purpose of taking into account the social and economic reality in accordance with capabilities and resources available and comprehensiveness in the strategic planning purpose of the briefing of all variables for production management and coordination. It is intended to include trends in the planning process during the successive periods of time either continuity intended to be an ongoing planning process leading to integration with the interrelations and participate either centralized strategic planning refers to the preparation and implementation will be decentralized.

Strategic planning is the planning of operations such as the direction supported to identify trends and direct paths solutions according to specific models and forecasts to measure the dimensions and standards of cost analysis and the development of substantive topics. And used to differentiate institutions and education strategic planning studies for the purpose of making educational policies.

The methods of strategic planning can be inferred, as well as methods of making educational decisions can be carried out through:
- Desk studies.
- Analysis of some studies of models and educational policies.

Analytical approach to strategic planning to guide paths solutions according to specific models that can be relied upon for the purpose of forecasting, as well as that depends in accordance with the strategic planning approach for the purpose of treatment and guidance along the ideological theories and using some multiple directions for the implementation of the curriculum, including information systems and methods of performance and conformity with the standards.

The pottery and the elements of strategic planning:

- Bank Information "Information data bank" (data, statistics and information).
- Realistic targets "Realstic goals" includes the ability to achieve goals and clarity of vision.
- The provision of experts "Providing experts" and contains a sufficient number of experts in planning.
- Availability of the laws and regulations of financial and administrative "Availability of the laws and regulations".

The analysis of the factors of strategic planning can be classified in educational institutions as follows:
- Strengths
  - Teachers and qualified teachers.
  - Integrated infrastructure.
  - Sophisticated equipment.
  - Modern curriculum.
- Vulnerability factors
  - The high cost of school supplies.
  - The weakness of educational services.
  - Deficiencies in the extra- curricular activities.
  - Weaknesses in the external communication.

Strategic planning in Iraq requires to enhance the properties of the future in the areas of social and professional knowledge and classroom through the dissemination of good morals, and develop citizenship, strengthening and developing the capacity for primary education.

The development of strategic planning in the field of education through the goals and visions, values and promote the role of environment in education, harmonization and simulation models actors with distinctive high quality and partnership institutional, community and the proper management of teaching and learning and improve legal and regulatory frameworks, support and develop the infrastructure facilities.

The importance of strategic planning for education to:

- Developing a framework to determine the future of the educational system.

- Encouraging the cosponsors of the education to work together and participation in the formulation of a common vision and standard of education.

- Clarity of vision of future goals and objectives for all beneficiaries and based on this vital sector.

- Opening the way for the participation of a wide range of diverse segments of society in the formulation of the strategy.

- Raising awareness of the importance of change and raise administrative efficiency to bring about change.

- Evaluating the previous phase of the survey of environmental destruction and to identify strengths and weaknesses in the educational system and the challenges it faces.

- Fruitful direction for the efforts and resources and invest them better.

- Strengthening the role of government and institutions involved in setting priorities according to a study of scientific methodology.

- Helping in devising ways and new mechanisms of action to improve their performance.

- Identifying areas of change and the challenges facing the education system and develop appropriate solutions to remedy them.

In implementing the Strategic Plan it require the identification of multiple programs and the formulation of accurate public targets for these priorities with targets procedural sub- levels differential activities and plan focuses on the solid ground of the operational levels in the provinces. There are a number of factors that support the implementation of the plan, including the role of public opinion and systems monitoring and evaluation of standards-based and continuous professional development and flexible plan and financial Amward.

Strategic priorities of education in Iraq

- Improve the quality of manpower and increasing the relevance according to the following mechanisms:
  - The admission policy.
  - The educational process.
  - Calendar and upgrading.
  - Infrastructure.
- Develop close ties between education and research institutions and the needs of the labor market.
- To achieve flexibility between manpower and the work done by.
- Priorities, particularly in some areas.

Raise the quality of manpower and increasing the relevance

This is supposed to form the main theme problems of education in Iraq through the need to influence in the following areas:

- Admissions policy
  - Enrollment rates (demand).
  - Standards spirited (response or request).
  - Distribution of the offer (map geographical and social) and equal opportunities.
- The educational process
  - The teaching staff (teachers, teachers).
  - Content of the curriculum.
  - Teaching methods.
  - Language of instruction.
  - Science education and technology.
- Evaluation
  - Evaluating the curriculum.
  - Evaluating the teaching methods.
  - Evaluating the collection.
- Infrastructure
  - Types of structures.
  - Temple of the horizontal (reference).
  - Vertical structure.
  - Education in traditional institutions.

- Distance education.
- Continuing education.
- Education Referrer.
- Rehabilitation.
- Self-education.
- The need to avoid the lack of human frameworks in informatics, communications and biotechnology.

Alternatives to the strategy

- There are a number of strategic alternatives for the development of education in Iraq
  - A strategy to continue writing.
  - A strategy of partial reform of adequacy of the Interior.
  - A strategy of partial reform of adequacy of Foreign Affairs.
  - A comprehensive renewal strategy.
- There is strategic alternatives for the dissemination of scientific and technological knowledge and to develop frameworks of human
  - Alternative rationing.
  - Alternative locational.
  - Alternative Technological transformation.

Strategic alternatives for the development of science and technology in the education sector

There are a number of strategic alternatives that offer a wide choice in the use of science and technology is focused on the potential offered by science and technology. Including alternative rationing continuity and reform alternative is the most important where it is updated, some aspects of the current school and educational planning and management, which depends on:

- Increase the use of modern educational methods and technologies in education.
- The introduction of new forms of technologies and vocational training.
- Reform of the curriculum.

Strategies
- Priorities
- Technical education and strategy
- A strategy for public education
- General goals of the strategy
- Strategic short and medium term
- The five- year strategy
- A proposed strategy
- Globalization and education

Priority

Many believe that addressing the challenges facing the education sector efforts are required at the highest levels (decision- makers, legislative and executive bodies) and thus the faith they need to Itenbwa the educational sector at the forefront of reconstruction and the adoption of strategies to deal with it.

That the status of the education sector a top priority is the right way to address the dilemmas faced by other sectors and the investment in education will contribute to give better results in many areas such as health, family planning and other services (water, electricity, nutrition). Spending on education contributes to the development of human resources and budget of the relationship between education and human development is very clear in countries with high level of human development the level of average spending on education 48%, compared to 4.2% of the countries with mid-level and 28% in the diameters of human development from low level Despite the significant increase for 2004 is still the proportion of funding for the education sector of the GMP, low education and will continue to share of gross domestic product "GDP" very little, if left education budget allocations for 2004, as is now would be the rate of spending per student $ 120, compared with more than $ 3000 in diameters advanced. That the education budget 2004 to contain any field of investment and reconstruction, only the costs of salaries and operating expenses of institutions and the Ministry of Education.

The purpose of the strategy

- Raising the level of basic education to become comparable with the standards followed in developed countries.
- Give high priority to increase Tnadea not possible in the demand for public education in the allocation of government resources.
- Rebuild the academic curriculum focusing on scientific disciplines such as mathematics, science, computer studies and principles of the economy.
- Developing teaching English since the beginning of the first year of basic education.
- The abolition of schools with meal or three meals.
- Encourage the private system to take a greater role in the education sector through appropriate measures.

Strategic training of manpower Purpose:
- Preparation of cadres from the various disciplines required for productive activities and the activities of science and culture.
- Highlights the human element competencies Kashkalip central theme.

Technical education and strategy

From the forces of change affecting the technical education, the growth in global markets, coupled with severe competition and the emergence of services and knowledge- based industries and the implications of information technology on society in general and the labor market institutions on the degree of respect and the changes in ways of organizing work within the institutions, for example, rearranging the administrative structures increased delegation of authority and emphasis on working as a team and multiple skills per capita and population change of Alnmwalscane and more and more young people aspiring to better jobs and social changes resulting in changes in the levels and lifestyles as well as the growing demand of consumers over a wide variety of services and at levels of high quality.

Lies the importance of the strategic outlook that takes into account the changes that occur rapidly in Manlv areas necessitate a fundamental shift in the methods used ownership of manpower, technical skills required and the challenge is to find creative and technical education and creative at least in the level of a technological education in the countries of the world, but and development and add to them as well.

Accordingly, when developing this strategy for technical education is supposed to be taken into account:

- Awareness of the role of education in achieving economic and social development.
- Raising the competitiveness of the country and support in the public and private sectors.
- To provide technical education and vocational training a modern and flexible to meet the requirements of the labor market of skilled labor and semi- skilled.

Strategic short and medium term

Steps

- Update the quality of data for planning and clear- cut.
- Restructuring of the Ministry of Education.
- Rebuild the infrastructure of the education system.
- Implementing a comprehensive program of training.
- Implement programs to raise the level of education (curriculum reform).

Five- year strategy

Principles of the Strategy

- Strengthening the performance and organizatonal and management capacity of central institutions.
- Adoption of the balance between centralization and decentralization.
- Planning incisive and modernization of administrations and financial functions.
- The establishment of systems performance evaluation.
- Reform the educational system.
- Raising the quality of education.
- Set priorities properly.

Proposed strategies

First: Admissions Figure who was responsible activity indicators / results

- Review the basis of acceptance and approval of the Council of the Ministry of Education- students reflect merit, equity and equal opportunities
  - Ensure the quality of graduates.
  - Reflecting the implementation of international cooperation agreements.
  - To achieve the greatest possible degree of harmonization between the wishes of students and disciplines available to them
- Determine the number of students admitted to the Council and the Ministry of Education.
- To accept foreign students and their distribution to educational institutions official board of the Ministry of Education.
- Admission of students receiving diplomas from non- Iraqi educational institutions.
- Admission of students in parallel programs and international and supplementary educational institutions and evening.
- To accept students in the disciplines of physical education and arts educational institutions.
- Acceptance of students in educational institutions, educational institutions.

Second: programs

Figure who was responsible activity indicators / results

- Review curricula of educational institutions and programs - programs to keep pace with scientific and technological developments and adapted to market requirements of science
  - Distinguish certain educational institutions in specific disciplines.
  - Develop the capacity of members of the teaching staff and develop methods of teaching and evaluating students.
  - Providing educational institutions qualified cadres.
- Perform an evaluation of programs in each university to select the best of them, strengthening and instructive cadres and equipment to become centers of excellence of educational institutions.
- Expansion of education programs of the Ministry of Education.
- The establishment of centers to develop the educational performance of members of the Ministry of Education.

- To send envoys to obtain a doctorate degree from a prestigious educational institutions in the disciplines required of educational institutions.

Third: Information Technology and Communications

Figure who was responsible activity indicators / results

- Update computer courses to fit with the output of high school in this area and the Ministry of Education - the increased likelihood of graduates to employment opportunities.

    - Improve the efficiency of teachers and graduates in the recruitment of information technology and communications.

    - Increase opportunities for communication and cooperation between members of the educational.

- Use of information and communication technology in all programs, including distance learning programs educational institutions.
- Develop the capacity of the teaching staff of educational institutions.
- To provide the infrastructure for the use of ICT educational institutions.
- Development of operational plans for the use of ICT in education and the Ministry of Education.

Chapter nine
Technology and scientific research

Preface

Technology is old and secured modern content associated with history of human and associated with developments in the twenty-first scientific century, but the technology content is evolving and renewed according to technological interpretations, and Greek documentation has contributed to the construction of the term technology, which consists of two syllables. The former (techno) means as mentioned workmanship and profession and the second (logy) refers to science and therefore complications goes back to the Greeks to consider technology as a science with the creation of new problematic in providing science technology).

Technology is linked to a science since it is related to its applied side, the technology must come on appropriate objective laws governing the movement of natural transformation and the evolution of man, and that the technology preceded the timetable science then became paralleled in modern times, and then the invention of the steam engine has occurred and its use in the beginning of the industrial revolution .

The aircraft has been used by the building and discovered the theories of aerodynamics, then diminished the time between scientific discoveries and technological applications, as a case of photography, and the status of the integrated circuit.

The technology is different from the science since it plays the role of mediator between the scientific centers and community, through transferring scientific findings to tools that meet the recent requirement , this role in the development process through the technology "the art of knowledge investment" to accommodate art all the creative capacities of individuals, universities, and technology is means controlled by the man to the outside world, the inventions and applications that are used in the production process group.

The obstacles of the technological advances summed up scientifically in the following features: -

- Chaos in the economic systems of the various sectors in the community.
- Illiteracy and lack of women's participation in the development process.
- The absence of equipments and scientific research centers, and the role of the media and technological publishing centres.
- Poor coordination between educational and training institutions and manufacturing establishments .
- The lack of central planning for the transfer of technology and ,development and localization.
- The lack of rehabilitation of technical and necessary engineering capabilities.
- The absence of national institutions capable of holding self-reliant technological development.
- Confusion in the selection of appropriate technology to local environmental conditions.

Technology has led to a radical change in the course of life so it is imperative to understand the nature of the changes in order to facilitate us to deal with the technology and make use of them properly on the basis of principles and away from the spontaneity.

Accordingly the technology can be defined, in the modern sense, as a system of experiences, skills and industries that lead to the provision of products or services based on the accumulation of knowledge and science and discoveries.

Mathematics flourished among the Arabs in the Islamic period in trigonometry equations, for example, has been finished to the laws of the foundations, the Muslim scholars has already been in use mathematical conclusion in the collection of some sequences and model of the sons of Musa Bin Shaker in their book tricks which we today call mechanics. To address the issues that remained for a long time is significant for the society and will fall in most of them the responsibility and efforts of the Organization of renewable critic undeveloped conscious mathematicians in this way.

The society issues would remove outlook processors that provide the mathematics grew up in the ivory tower and the productive activity of man. The study of the history of science, history and the history of mathematics describes the prosperity of civilizations that has been linked to the prosperity of the mathematical sciences.

The event in 1996 was exciting, the birth of Dolly in a somatic cell into a specialized egg-enriched after removing core and planting it in the womb; the most important point of this event is the return of specialized cell and embryonic stable situation after losing this status. The other development in science is the production of sufficient quantities of food in the world. Many thinkers expect that the world will see a lot of problems, related to scarcity of resources and energy, such as increased pollution and population explosion. Most studies of future ending 2025 required further means such as:

- Technical means (computers to store and recall information).
- The use of special programs to make the prediction.
- Many experts, technicians and programmers.

Possible divisions of future studies in science include three types; this division is used haphazard to simplify them, as it is difficult to separate the three types of studies, from each other, such as:

- Studies that rely on prediction (what will happen in certain area of science).
- The overall outlook studies that rely on intuition which looks at the impact of current scientific achievements on the future of humanity.
- The overall outlook studies relied on detailed statistical information within the mathematicians programs on computer models.

Frontier sciences

Academic disciplines have evolved with the development of sciences and various new disciplines such as engineering, agriculture, science and total treatments that began in the nineteenth century, whereas the twentieth century indicates other disciplines such as business management, journalism, information and library science, economics, politics and world affairs were added.   Each state has its own special methods to determine its own disciplinary university and identification numbers, graduate students and the quality.

The world witnessed in the twentieth century breakthrough in all fields and scientific trends, so there are no boundaries between different disciplines. For example, medical science requires engineering science and recent tests of modern science depends on the physical, chemical extraction and analysis and also relies on mathematics to lay the groundwork mathematics.

The progress and development in pure science, for example, develop new subjects and disciplines and specialties of science. New interfaces were not known during the first half of the last century. The results of these major changes in curriculum and build up research transformed these developments to the university curricula. Seminars and researches are now carried out in different ways including:

- Bachelor based on the study and theses.
- Some universities in Britain and Germany developed curricula at the level of initial studies that include research and study.
- Divide the present fields such as industrial chemistry, chemistry of life with medical side and other disciplines in the branches of pure science.
- Develop competencies.

These terms of reference have been developed in American universities in physics, chemistry and mathematics, to prepare graduate in some sectors such as engineering, chemistry physics and chemistry, agricultural engineering, and medical studies.

- Adding assistance topics.

Some topics have been added as assistance of many of the terms of reference of pure sciences, including education, literature and library services and use modern machinery and computers.

- Other disciplines (Sandwich)

British universities were carrying out by expanding the initial years of university study for use in increasing opportunities for the systematic teaching and applied for and rehabilitation work in various production sector.

Techniques used in science
Diagnostic Imaging

In medical diagnosis it is adopted mainly on the knowledge of diagnostic imaging technology spectrum, including the use of X-rays and gamma rays from, which is characterized by being electromagnetic radiation ionizing radiation, then began to think about using the term of this non- ionizing radiation infrared or microwave radiation and technical NMR magnet. The examples of spectral techniques used in diagnostic imaging:

- X- rays
- Gamma- ray
- Ultrasound
- Infrared
- Anti electric tissue
- Visual mechanisms

X-ray

The oldest techniques that is used in diagnosis and therefore will not focus on the importance of being where they were getting on the first pictorial representation of various tissues obtain after the development that is built on a limited computer assistance.

Gamma rays

The purpose of gamma- ray is the imaging profile then it was developed as computer- assisted also in the eighties which was called "ECT" and was then developed using imaging "Postiron emission tomoyraphy (PET)" where the radiation of tissue is carried out by position (positively charged) and thus can get a picture to clarify the life processes of the tissues that carry electrons and draw.

Ultrasonography

The speed of these waves are characterized by being less of electromagnetic waves, which provides an opportunity to measure the fetus as well as during the stages of development in the womb, added to that the fact that this technique is based on the fact that the X- ray is not ionized therefore it is not a preferred use in diagnostic imaging.

Nuclear magnetic resonance imaging

Despite this technology it is old, but it was then developed for the purpose of medical diagnostic imaging has gone from the seventies, where the nuclei of atoms is measured by the disposal of certain substances found in different body tissues. The criterion for the disposal of these seizures depends on the radio pulses that are similar to the frequency in the field of outer- core magnet and thus to obtain a diagnostic can be used.

In the medical applications for the purification of nuclear magnetic resonance imaging to obtain imagery of infarction that occurs in some parts of the brain and important developments in this area the integration of multiple techniques and access to advanced apparatus for nuclear resonance imaging, including the "TMR" and "MRI".

It is important experiments that experiments are used the magnet resonance imaging of kidney transplantation, which was filmed nearby parts of the kidney and then infected the interactions that take place within the body after transplantation and efficiency of the cultivated parts. As well as imaging of tumors within the liver and liver imaging at the time of myocardial fibrosis or within, as possible, filming parts of the stomach and colon and to identify tumors. It was also to obtain information about stroke and is believed to imagery obtained of cancerous tumors of the brain were more pronounced than the use of X- ray.

It can be measured by any inflation occurs as a result of heart disease, and can also study the problems of the heart due to the presence of any obstruction or infarction in one of the blood vessels and could also portray the evolution of stroke, heart attack and its impact on the heart.

A nuclear magnetic resonance imaging "MRI"

This device is used which was created as a result of the development in the technology of magnetic resonance spectrum by the registration of spectra of life processes taking place within the animal body where the magnet-making with full slot by placing the human within the magnet and thus these devices provide a complete picture of the part which is conceived, and the advantage of the fact that this device magnetic field is not harmful, and the microwave radiation used is not harmful too.

It is possible through this device to study the effects of ongoing parts of the human body while taking a particular medicine can also be follow-up of the various core elements and sequentially, as well as to study the changes occurring stereoisomers of chemicals inside the cell as a result with other molecules.

New Technologies Electron Microscope

Electron microscope is using a torrent or stream of electrons, where the wavelength is too shoat then we can get on the ability of the analysis is very high. Extent of segregation (analysis) of the optical microscope and is an estimate of 2000 and this is not enough to see parts of the cell, viruses, and macroscopic particles, but the use of electron microscope segregation less than the uranium atom (in approximate) in special circumstances. It is clear that electron microscopes large and complex and expensive operation that is similar to the foundations of the optical microscope, which reveal the sale electrons emanating from the source mail (metallic thread with a high degree preheated in vacuum) and reveals extensive by lenses and lens-body grows electromagnetic diffraction and finally drop the image as by the final lens of the projector.

To see the image on the screen is up brilliantly by the lens or can be scanned to imagine, as we mentioned earlier, the high segregation ability of this microscope to enable the researcher to view more details when you enlarge the optical microscope, the exact address. When the materials to be examined too thick for the passage of electrons then therefore requires the creation of a thin section, and it must be the sample used for the purpose of this solid and cut easily.

Labeling with radioactivity

Require a lot of chemical analysis revealed small amounts of material with amount of concentrations $10^{-4} - 10^{-6}$ molari therefore it requires the development of other ways to respond to the concentration of low- lying, such as the development of experimental methods by radioactive to solve many of the other problems that might face them. Some of these methods that could be used by dual- labeling for follow-up of two similar materials formed at various times by pulse method for follow- up fugitive substance at a time after the configuration without interference of other material. An example is the use of radioactive materials in the chemistry of life:

- Choose a material that resides on small concentrations, which are difficult to measure by direct chemical methods.
- Distinguishing similar molecules in different chemical sites.
- Analysis of mixtures that are very complex, which can not be done by various conventional chemical methods. Including:
  – Enzyme interactions (DNA polymerase).
  – Measurement of molecular weight of the DNA by labeling the final group.
  – Diagnosis particle by settling with the anti body.
  – Protein purification, which does not have a chemical test.
  – Diagnosis of active centers of enzymes.

Autoradiography

This method is used to detect and locate radioactive materials in the cells or tissue for example, and so the molecule itself and is done by the impact of radiation emanating from radioactive materials or emulsions of photographic plates specially designed for radiation imaging device self-motivate, where silver halides grains, located in the emulsion as a result of the dissolution of radioactive materials in the sample, and the emission of radiation, including activation and work output reduction as indicators minutes for the site radiological effectiveness. And signaling models resulting from the grains chemically and radiation efficiency in the presence of structures that are in contact with these granules and the microscope can be obtained from the resulting image on the two types of information at the site of radioactive materials and the quantity of a radiation of as the amount of silver particles is directly proportional to the severity of radiation present.

Of the modern applications of this technology as follows:
- Measurement the number of molecules of DNA in bacteria phage.
- Measurement the number of secondary units of the chromosomes.
- Double vision in the DNA molecule of bacteria.

Membrane filtration and screening "Membrane Filtration and dialysis"

(Like cheesecloth that has been used to separate the serum from the leaky). The cheesecloth filter extracts used in textiles. Then use the cards instead of porous fabrics for the purpose of controlling the size of chips and then create filters made up of cells, or glass yarn, either the softest materials they include sorting tubes membrane, which allows the passage of small molecules and ions. But keep the particles and macroscopic aggregates macroscopic particles. It is called the membrane tubes stitches the contrary, and that have the ability to separate the macroscopic particles from small.

Protein engineering

It is the technique that allows the installation of structural proteins desired in order to build a clone- mediated DNA "Cloned DNA". There is no relationship between the latter and engineering of proteins used, including the building of protein functionally, chemically and physically.

The DNA could be modified by two ways using:
- Mutagenic in private venues.
- Switch sections of the nucleotides.

The protein engineering include modify the structure with protein mediated by genetic engineering and most protein engineering is carried out currently in the field of enzymes, either to speed up its response to the incentive or to become more receptive to acid and heat.

Example: "Cloning" the cDNA for the receptor of "acetyl choline receptor" facilitated the technology which is called site directed mutagensis for getting sequences skilled "Deletions" or substituting some of the amino acids in an additional unit "subunits" of the receptor and then it can test these changes on the functional aspect, and are also defined as follows:

There are many examples of this type of modification for production of complex of organic compound that have catalytic activity have of it chemically synthesized for example the myoglobin of which associated with oxygen, but it docs not have catalytic activity. This Bio- molecule with three complexes of ruthenium "ruthenium" carrier of the electron through the surface of the histidines components generate a complex that has the ability to reduce oxygen and the oxidation of the natural ascorbate.

The construction of DNA contributed significantly to the development to the stage of protein engineering to construct proteins that do not exist in nature. The technique has evolved to the point can modifies the gene by an

engineering to change the protein in a predictable and have to improve some functional characteristics such as:
- No. transformation "turnover number".
- Static Km of substrate specific.
- Thermostability.
- temperature optimum.
- Stability and activity in non-aqueous solvents.
- Privacy of interaction and substrate "Specificity".
- Requirements of co- factors.
- Protease resistance.
- Allosteric regulation.
- Molecular weight and composition of the structural unit "Sub- unit structure".

And for engineering the protein molecule, it is clearly necessary to ensure a series of rules relating to major synthetic building blocks of proteins that recipe as desired. After seeing the structural composition of protein crystals, it is then possible to diagnose those areas in which it occurs possible modifications to improve the catalytic molecule, protein, and this is done to modify the sequence of amino acids in the protein.

Major modifications protein

The use of site- directed mutagenesis determined then what is aimed to, because the change in one base in the gene result in a change in the sequence of amino acids in the protein, which in turn improve the protein in question. Large modifications in proteins by removing the "delete" section mediated by enzymes or by the unequivocal chemical structure of part of the gene. In this way, the production of spare "klenow fragment" "DNA polymerase" free of analytical activity, also can add sequence of amino acids through docking to improve the stability of proteins made in E. coli and finally can collect or part of a fusion gene or the whole of all or part of the other, thereby generating new proteins.

Determination the general features of the installation of the structural protein. Protein engineering based on the availability of information on the district and synthetic building blocks that are obtained from the methods of

X-ray diffraction and nuclear magnetic resonance two-way "Two dimensiona nuclear magnetic resonance NMR" and the latter is the alternative method in the future. Many researchers expect success in engineering of proteins "Protein Engineering", especially after the great progress which has been in embryonic technique, where each protein is produced by genetic conditions of its own machine of the cell consisting of enzymes when they become three characters of the genetic material and arranged in advance and checked that then wrap as a specimen to be specific proteins effectively.

When you know the rules that allow the protein to form belts wrapped can then change the genetic information of proteins and identified so that it works in another way as soon as a large and powerful grants stability, and thus can benefit economically from the proteins of the broad areas of application by micro-organisms and can be more clear: for example, improved production of proteins (new physical properties and functional).

Important notes that are related to protein engineering is to clarify the potential relationship of proteins, where the protein for example, a specimen 15-amino acid. There are 103×3 possible sequence of these acids is larger than the number of atoms that make up technical enzymes immibolized onboard, the development of these enzymes are restricted or limited to a solid surface to be in constant contact with the foundation to which the article in the mobile phase "mobile phase". It is clear from this that there is a possibility to use the many pathways that retains its effectiveness.

Technical features of immibolized enzymes"
- Prevent the entry of the immibolized enzyme in the mobile phase.
- The product is characterized by being cleansed of the enzyme and does not accumulate.
- Using the enzyme for long.

The globe, despite the lack of clear understanding of the rules that govern protein engineering, but the equipment contribute to give some suggestions on how to achieve a stereo structure of the protein. In this area one can not expect for example bacterial cell to produce human that differs in form of human protein.

Immobilibzed enzyme technology

At present, there are important industrial applications of immibolized enzyme technology represented by the following enzymes:

- Glucose isomerase.
- Aminoacylase.
- Penicillin acylase.
- Lactase.

The latter has been "Immobilized" on the particles of silica. It is used to convert the lactose in whey to glucose and galactose.

Applications to include of immibolized in the future as follows:

- Use enzyme "Cholinesterase" for the purpose of pesticide detection "Pesticides" and watching the inhibition of this enzyme either by the method of electrical "Calorimetrically electrochemical" or by the color method.
- Other enzymes that may be used in the same method in order to detect toxic chemicals, the enzyme "Carbonic anhydrase" is very sensitive to low concentrations of chlorinated hydrocarbons from low- lying "Chlorinated hydrocarbon" and "Hexokinase" to "Chlordane".
- Immobilized diisopropyl phosphor fluoridate extracted from the nerve cells.

General aspects of enzymes immibolization

This process is intended as we mention it to determine kinetics of enzymes, as yell as cells that characterized by (desorption) on the surface such as fibers gels, etc., also can be used as phenomenon shooting accordingly.

Advantages of the immiblization process are the followings:

- Finding the status of enzymes similar to those found within cells and tissues.
- Prolonging the period of use and has repeatedly given to the survival of catalytic activity and stability.
- Use appropriate concentrations and may be high for the purpose of increasing the speed of the reaction, given the focus to fit with the speed in specific circumstances of the reaction.

- Contributing of the immibolization process to facilitate the purification process of related to products of reaction.
- The use of multiple systems from the fermentation (continuous and open).
- Reducing energy consumption and cost.

The immibolization methods are numerous, including:
- Chemical methods: they are similar to affinity chromatography such as use the covalent and casual.
- Physical methods: such as packaging inside a capsule adsorption and shooting.

As for choosing the appropriate method to be immibolized are determined according to the specific bases represented by measurement of activity, stability, so it must be taken into account the business side that is, have used with less expensive. And choose the easiest method because they are all tough and stay away from hazardous substances to human health, and the technical side is important in the selection process since there is a special mechanical pressure during the operation.

The immibolization cells vary from cell since it is being more of enzymatic system builders with the installation of diverse chemical content, therefore, requires that the appropriate modalities, simple and stay away from these that require to use extreme circumstances. It also requires that to taken into account the number of cells to be immibolized so the method must be a convenient and linking cells are good and avoid the use of hazardous materials. The characters of the immibolized cells are numerous advantages including the use of small amounts of carbon and energy sources and re- use of cells, so it is possible separate the growth phase from production phase, where it is possible control the fermentation. Immibolization depends on the type of cells, microbial cell reduce the size of the manufacturing process and thus reduce the cost of the production process. The Eukaryotic cells which are characterized as specialized capable of limited division of which are specific plant or animal cells and preferred to be immibolized, particularly those that are separated as any single and are generally used for the purpose of the immiboliaztion of adsorbed on the hollow fiber.

Enzyme Technology

The biotechnology is considered as one of the technical life in science and engineering. It was one of the enzymatic technology trends that have grown with the technology of life, despite being preceded by technical life, keeping in mind that enzymes from an engineering standpoint is a special case of the factors that have qualities such as privacy.

Bio- systems are used in critical periods in history to get the desired chemical conversions such as transformations of like milk to cheese and fermenting of liquids that contain sugar to alcohol, but such research trends have changed during the evolution of Biotechnology with the fact that these processes such as cheese, bread and alcohol industry still very important.

The history of enzymatic techniques started with the developments that have emerged a number of chemical transformations using the tissue of life, which include, for example hydrogen peroxide decomposition and degradation of starch to sugar and digestion of proteins.

Human Genome Project

The initiative of (HGP) came the first time in 1988 and aimed at finding the sites of some 100000 human gene in DNA and the (HGP) expresses 24 pairs of the human of chromosomes, is turning into information content when it follow the sequence of rules need to be resolved based on computer science, mathematics, statistics and experimental sciences. Note that the computer science often provided in contributions in programs and solutions that are characterized by the skills that led to the invention of language access code information described the performs a particular order and provides methods to describe complex biological processes by the number of code rather than their natural language with hundreds of pages. Then the researchers and observers said that the twentieth century be the century of biology and analytical power resulting from the HGP that will explain drastically all life and medical research as it was:

- Research on the nature of genomes and the nature of the composition and organization of various scientific institutions.

- Acceleration in the implementation of the project, which was planned to complete within 15 years of technical progress, but then shortened the time to ten years.
- The search of the genome to find the type of information or material contained in the communication as well as identifying the sequence about three billion chemical bases.
- Study the nature of the information stored in the computer and the evolution of elaborate by efficient techniques of and sequence evolution in the tools that contribute to the analysis of information.
- Study the effects to be set in the community and to what extent this can be achieved, and the type of response.

In spite of all reported studies and research conducted by methods and techniques in various vital information as well as numerous writings and published in this area, there are still many other fields and various study and research. Some of these fields has not been touched so far in the country, especially the human genome projects, and areas to attract the attention of researchers, but it's mostly a few problems, mostly dealing with partial or subsidiary.

Bio- engineering

There are a number of scientific developments resulted from the diving in the world of molecules to push medicine forward through the discovery of technical of recombinant DNA (engineering life) and this new knowledge has led to the understanding of the causes of the disease that has eluded science until now, and thus to find new treatments to them. Engineering of life had an impact on medicine borders these have become easier with the forgotten youth of this important scientific field. The reality is that James Watson and Francis Crick did not reach a structural installation with a double helix molecule of DNA. And then it was identified the gene (genes), which manages the production of individual proteins, and then we obtained the tools of partial strong, and in the early seventies researchers began snapped genes of the DNA. One of the species and planting it in DNA another kind for the manufacture of new molecules and in a few years researchers were able to transfer these genes and to produce objects that are within during the eighties and became a human gene transfer to many microscopic organisms and bacteria turning them into factories for medically useful proteins.

After it has been cloned of human genes in the micro- organisms for a number of hormones, including growth hormones and insulin in human as well as bacteria many of the genes responsible for human proteins with diagnostic value was produced at the level of marketing. It is noteworthy that human insulin is derived from living with diabetes, and also for the development of techniques for the production of antibodies "monoclonal antibodies". Many applications, there is a steady increase in the use of enzymes in the diagnosis and treatment as well as in planting (farming) tissues and cells, "Tissue and cell transplantation" and that the development of engineering of life is still in the young stage, but there have major impacts on medicine and industry is synergy between electronic systems, electrical and life- component electrons so- called life "Bioelectronics" and electrochemistry of life "Bioelectrochemistry". Then there have been the following design of a number of devices depending on what is stated in the above examples include "Glucose monitors" for the purposes of medical sensors and nerve gases for medical purposes and sensors nerve gases "Nerve gas sensors" to military uses. Based on sensors that have been most developed in the present time to reveal the exact products enzymatic activity mediated by the traditional pole "Conventional" where is the install (restricted) "Immobilization" new approaches that lead to devices with more sensitivity that depends on the movement of electrons between the direct-polarization and the redox centers protein "Protein redox centers" In brief the enzymes, which is based on the sensor depend on the medical sensor "Glucose sensor" and other sensors that measure chemicals in blood such as immune sensors include the electronic life "Bioelectronic immunosensors", which was commercially manufactured during the current decade, are measured in a large number of materials in the fluid of life, causing a revolution in the diagnosis, in addition to the incremental progress that has been happening as a result the development of a wide range of models "Sensors", which depends on the synergy between micro- organisms substantiated grants stability, "Immobilized and Stabilized". Finally, various data indicate that the microbiology of life through the engineering involved in the medical field in the production:

- Antibiotics.
- Vitamins.
- Nucleotides.
- Hormones.
- Enzymes.
- Vaccines.
- Antibodies.

The progress that accompanied the engineering of life has affected in particular the daily practice of doctors, because of the speed that accompanied the evolution of knowledge and techniques in the laboratory and hence to the industrial production and then patient care. The expression of human insulin gene in bacteria E. coli, for example, has been studied in 1979 and that this insulin, with the original engineering- life of "Recombinant DNA" has been tested by volunteers with non- diabetes "non-diabetic" in 1950 and clinical trials that have been in patients with diabetes began in 1981.

The attention of most doctors on the applications of modern life engineering in medicine, which tend to be very important in areas which have helped to revolutionize the diagnosis, treatment and understanding of many diseases, and examples of this therapeutically important protein, which was manufactured by engineered mediated microbiologist, microbiology, applications of single origin "Monoclonal antibodies", enzymes and others that arise out of uniform origin from lymphoid cells, where used in:

- Treatment of cancer.
- Diagnosis of many diseases.

Pharmaceutical industry, pharmaceutical companies have been choosing some clinically significant produced cheaply, such as insulin, which was previously mentioned, and which treats patients with diabetes and extensive use of interferon for the treatment of many diseases, including cancer.

The Bio- engineering worked towards a second method by increasing the secretion of microbiology by called anti- life penicillin produced in fungi, and the third trend in the medical field that is the development of drugs already in the nature and turn them into centers of drugs more effectively.

Containing anti-bacterial drugs that have contributed to engineering life and developed vaccines, hormones, vitamins and antibiotics and life for the purpose of producing these materials from micro-organisms after it was restricted to human and animal cells.

Hormones are the most advanced in terms of the accuracy of the technique used and the large economic returns through the engineering of life and led to great successes through the production of materials likes of the hormones which are stimulated, and stimulating the flesh wounds and the growth of the affected nerves that affect the sense of pain. The success of engineering in the provision of life- hormones of the study and treatment has been a boom due to technical difficulties in extraction, which vaccine and growth hormones as well as the instigator of the secretion of pituitary adreno "ACTH" used to treat infections and diseases is used to treat wounds, burns, and stunting and release thyroid hormones pituitary as well as insulin used to treat diabetes, where possible transmission of their genes to bacteria.

Production of hormones is mediated by microbiology research center in the fields of engineering life in general and genetic engineering in particular, where microbiologists used to convert steroids and the production of hormones from the human body can not produce in sufficient quantities. Then it was grown in importance after the custom of cortisone and its derivatives and their effective role in the treatment of arthritis, which draw many medical companies of steroids from plants, animals and chemical methods of trying to turn them into other steroid prescriptions. The methods of microbiology steroids is turning quickly but with less degree, there is in the addition of specialized microorganisms capable converting steroids quickly. There is also the addition of specialized microorganisms is added hydroxyl group of any carbon atom present in the steroid. There are also some working to add hydrogen to steroids or withdrawal of hydrogen or oxidation or separate pools of chemical side effects. Using growth hormone that is released from the pituitary gland for the treatment of dwarfism find the hormone extracted from the animals be in a non-pure from, but according the production of this hormone is preferred to be extrated from microbiology such as the production from the bacteria E. coli after treatment genetically.

The plant hormones have been possible to produce from fungi, especially those produced from rice, as it is known that plant hormones industry is still expensive despite their limitations. In addition, there are a large number of proteins found in the blood such as the factors that contribute to coagulation missing by patients with haemorrhage as well as the albumin found in serum. These materials have been contributed to the development of production by engineering life in medicine (drugs).

The pharmaceutical industry, which includes anti- bacterial drugs, vitamins, vaccines and hormones of the biggest industries that relied on engineering techniques of life for the purpose of producing these materials from microbiology.

Medical applications of Bio- engineering

There are many faces, can be addressed when studying the medical applications of bio- engineering after the gene was designed, including:

- Production of therapeutic: include hormones, such as somatostatin insulin, interferon and anti- biotic, where it was initially isolate the hormone somatostain for regulating secretion of growth hormone from the pituitary gland in the traditional way that requires half a million sheep brains to produce 5-10 mg of this material.

- Treatment many of the genetic diseases: the treatment of many genetic diseases possible to treat many genetic diseases due to loss of protein production remedying these proteins from bacteria, and the examples of this case the planting and production of large amounts of genes to produce hemoglobin, which decreases in "Thalassemia" through the introduction of genes responsible for hemoglobin the patient's bone marrow, and then returned the cells to the patient.

- Diagnosis of a number of diseases before birth: the fetus diagnosed in the prenatal stage, through the identifying the defects in a specific gene that causes the disease, such as some "Gamma- Globuinemia" and the disease lest Nhin as well as Tay- Sachs "Tay- Sachs".

- There has been progress in some areas of medical engineering technology due to the recombinant DNA such as "cloning" the human insulin gene as well as growth hormone and its expression in bacteria that has been marketing of human insulin derived from microbiology and used for the treatment of patients with diabetes in addition to:

    - Production of interferon by a large clone human genes in microorganisms.

- The development of production techniques and monoclonal antibodies and their uses.
- The increase in the use of enzymes for the diagnosis and treatment in instilling the cells and tissues "tissue and cell transplantation".
- Treatment of many diseases of genetic mediation by protein that being lost, which can be mediated by production of bacteria.
- Diagnosis of diseases before birth by identifying the defect in a gene or several genes.

Turning to the relationship between engineering, medicine, is taken into account the following things:

- Mutant cells and the cells unmodified organisms and their products such as antibiotic cellular life and plants, as well as other life transitions "Bioconversions".
- Modified cells "Modified cells" and their products to ensure that objects Monoclonal "Monoclonal antibodies" of the following uses:
  - Immunological Studies.
  - Immunohistochemistry.
  - Tissue typing for trans- plantation.
  - Diagnosis and monitoring of malignancy.
  - Preparation of medicinal products with a "Prepartion of medically important products".
- Recombinant DNA technology and its use for the production of insulin, interferon and growth hormone and vaccines "Vaccines" and enzymes.
- The application of Bio- engineering techniques of molecular genetics and techniques diagnosis recombinant DNA in the diagnosis and (pathological) human disease:
  - Patriarchal diagnosis of genetic diseases.
  - Effects of genetic diseases on the specie disease.
- Features of the future.

It is believed to that Bio- engineering represented by "Clinical biotechnology" has begun in the application management and industrial production of penicillin in 1940 that the success of the full insulin has created a growing demand for medicine (drugs).

The production of penicillin by fermentation and used in the treatment of diseases using the Bio- engineering problems that has been accompanied by the emergence of side effects and put some Bio- engineering solutions , and the problem of production has been developed through genetic improvement producing strains and control the components of the center other conditions contribute to the process of fermentation.

Bio- engineering and cancer

Bio- engineering has succeeded results in the field of cancer better than other diseases, as shown in the eighties that the main thing vs. cancer is a change in the genes (genes) from an engineering standpoint.

It was clear from the following entries in the relationship between the Bio- engineering and cancer.

- Through analysis of a group of viruses called regressive "Retroviruses", which cause cancer in animals, as a number of these viruses carrying cancer- causing genes or tumor genes "Oncogenes". It appears that the retroviruses that cause cancer have been captured from the normal gene, cell, and one animal and made it part of their own genetic material. The retroviral infection of new cells in the later planted with genetic material, leading to the transformation of healthy cells into cancerous cells.

- The researchers show that DNA extracted from human tumors can shift the cancer cells to cancer cells in test tubes. Or that a specific gene in a human cell that can transform sound cell into the tumor cell and a tumor-causing gene for bladder cancer in humans and called "ras" almost identical to the viral gene, a causing tumors in mice.

- The gene tumor is often due to the mutant or increase in production and there is general consensus about the fact that any of the original tumor gene mutations may be some inherited mutations. The studies of funmor contribute to inherited breast or ovarian caner, the physician may be able to use that gene to assess the patient's condition and prospects and to provide more effective treatment for patients who have multiple copies of inherited suspicious. Harold Varmus and Michael Bishop has concluded that "Lancogen" the legacies of the genes responsible for causing cancer.

Bio- Engineering and AIDS

To understand the relationship between Bio- engineering and the AIDS requires a study of the topic in two cases:

- How should the immune system to destroy virus: the defense forces resulting from the immune system to attack multi- directional and of different media for the virus (a specific target) to:

  - Phagocyte and other cells relevant to specific viral antibodies are chewing.

  - These cells installed in the grooves on proteins known as antigens of human white blood cells.

  - Construction immune complexes on the surface of cells identified by a type of white blood cells (T-help) "Helper T".

  - The recipients are on the T- cells help identify the peptide superficial "epitopr", associated with divide, and secrete small proteins that stimulate and activate T-cells and the toxic or lethal trait.

  - The killer T cells directly attack infected cells and fragmentation of viral particles and peptides associated with molecules of antigens of human white blood cells, when identified by toxic T cells by antigenic recipients on the surface of infected cells and destroy them by producing more of them.

  - The B- cells recognize the antigen norepinephrine viral surfaces as a prelude to their destruction.

  - Immune response and the virus "HIV" contribute the immune steps in defense against the virus "HIV", where they are:

    - Invasion of the virus of T- lymphocytes and cells assistance, followed by cloning and increase the virus and help decrease the number of cells, death, and loss of infected T cells.

    - Launch of viral particles from the cell membrane of T cells after being wounded by the T cells and B- toxic responses to be dispatched a strong defense which resulted in killing infected cells, viruses, and thus is determined by the breeding assistance and reference cells to a normal level.

    - A high level of virus gradually with the decline in the number of cells to help patients and reflects the so- called phase of AIDS when the number of cells less than 150 assistance cell in the blood followed by a rise in the level of virus with the decline of the immune system.

Monoclonal antibodies

The areas of application for the production of these antibodies where the potential for many therapeutic and diagnostic enormous, including:

- Treatment of patients with leukemia and production of specific antibody alien objects on the cancerous blood cells, leading to the union of antibodies with and removed from the bloodstream.
- Accepting the objects of a transplanted organ which are used Monoclonal antibodies or clone in the development of the body accept a transplanted organ such as the kidney.
- Birth control through private industry specific antibody to proteins found in human sperm.
- Determining the sex of the fetus through a special antibody to sperm of own unwanted sex.
- Models are highly sensitive and privacy are being used as opposites, and a single origin and widely high sensitivity and privacy in early screening for malignant tumors by using specific proteins associated antigen and the presence of tumor presence.
- Determining the levels of hormones in the body and used Monoclonal antibodies to determine the levels of hormones in the body and determine the effectiveness of the glands.
- Search for the presence of some drugs in the body tissue and blood used Monoclonal antibodies in the search for the presence of some drugs in the body tissue and blood to prevent the occurrence of cases of poisoning or addiction.
- Diagnosis of crimes using Monoclonal antibodies in the search also in the diagnosis of crimes. The food industry also used Monoclonal antibodies in the field of food industries, especially in the diagnosis and determination of the purity of food, processed meat, and free of unwanted substances and preventing fraud in this area.

Of the significant developments that have taken place for Immunology and molecular biology and biochemistry and the discovery of antibodies and the creation of a single origin "The Monoclonal antibodies" is characterized by privacy "Specificity" and sustainability of production, "Immortality" huge quantities "Large ruantities" and high purity "High Purity" for periods of a very long time.

However, these antibodies Monoclonal antibodies created by the multiple origin (clone) the molecular composition and effectiveness. Studies have shown that the use and applications of antibodies only be successful to detect very small quantities of tumor functions that can be used in early diagnosis of many tumors and by diagnosing the effectiveness of these antibodies could be argued that a large proportion of blood diseases can be categorized.

The advantage of imaging the immune flashlight as we have mentioned that the blue single antibodies prepared in the body of a patient associated antigen, surface of cancer cells without other cells and sputtering when labeling these antibodies with radioactive isotope, it can locate the radioactive iodine, for example by gamma cameras and thus can be located and the size of cancerous tumors, including colon, ovarian and skin cancer. The unilateral clone in addressing some of the tumors where it can be linked to medicine as well as radioactive materials to these antibodies, such as chronic leukemia and thyroid cancer lymphoma and colon cancer has been found that these antibodies injected intravenously is grappling with the tumor cells and selectively and is disposed of, where became can direct these drugs directly to tumors by linking them to the catalytic antibodies to these tumors. Used monoclonal antibodies to treat cancer when there is a high toxic concentrations in the tumor. It could also be linked Monoclonal antibodies radioactive isotope and alive in the body of a cancer patient at which time the radioactive material to the site of the tumor and therefore within the cancer cells and it crashed. There are many researches addressing the use of monoclonal antibodies in the early diagnosis of the body rejecting the case of the tissues and the transplanted organs as well as a lot of studies on the use of these antibodies in the treatment of the case of rejection.

Some applications objects Monoclonal

Improving the sensitivity of the current immune for tests or tests new Histocompatibility

Fibronnectin

Blood groups Antigens

Sperm antigen

Interleukins IL

Interferons

Progesterone gastrin

Blood clotting factors

Estrogen

Human growth hormone

Monoclonal antibodies has clear impact and important role in clinical medicine before developing the "hybridoma technology", which provides heterogeneous objects "Homogenous antibodies". The research carried out by each of the "Kohler & Milstein" in the early seventies, created a method used for the manufacture of the anti body homogenized with a quantity of non-specific proliferation applied at large.

The researchers "Kohler" and Mlesstin have participated in the production of monoclonal antibodies, which is derived from specific tissue culture which is called hybridoma, where the latter's has the ability to produce one type of antibodies but does not produce more. This is done by crossbreeding or mating types of cells, the first is produce the antibody and the second for the growth of cancer cells have the ability to reproduce. And then treated with hybrid that has to be the formation of antibodies, where antibodies are produced for this body alone, and perhaps it carries the qualities of cancer, the production of antibodies is very large quantities. It is possible in the light of the use of a composition for the manufacture of an unknown antigen monoclonal each part, and then used these antibodies to probe the chemical composition of the real knowledge of the unknown substance.

Monoclonal antibody can be used for treating patients with cancer of the blood through the manufacture of these antibodies is specific to the alien objects on the cancerous blood cells united for the purpose of removal from the blood stream, and used these antibodies for early detection of the presence of tumor cells through the tests that require purity too high to measure the presence of proteins associated with its existence of these tumors and their locations in particular antigen- mediated tumor.

These antibodies are used in determining the levels of hormones in the body and to determine the endocrine events are also used in the search for the presence of certain drugs in the blood and tissues because of the poisoning have also been introduced in the diagnosis of bacteria in the development of the transfer of the body of a transplanted organ, in particular kidney.

Hot issues

These issues raised by biotechnology as a result they have become at the forefront of basic research and applied consistently reached new levels of progress and complexity and can be far- reaching impact and positions that require scientific, political, moral and social. These issues vary with varying impact and could be referred to some of them:

- Cloning.
- Human Genome.
- Gene therapy.
- Map of the protein.
- Food and genetically modified organisms.
- Advanced technologies (nanotechnology).
- Vital information.
- Monoclonal antibodies.
- Biotechnology, medicine, agriculture.
- The discovery of disease- causing genes.
- Forensic – DNA, Fingerprinting.
- Biotechnology and biosafety.
- Biotechnology and environmental balance.
- Scientific strategies of biotechnology.
- Research and development and stops future.
- The role of education and training in biotechnology development.
- The twenty- first century is the century of Biotechnology vitality and prospects and challenges.
- Technical aspects of genetic engineering.
- Bioinformatics

It can focus on hot issues of the following:
- Cloning.
- Human genome.
- Gene therapy.
- Genetically modified food.
- Ethics.
- Genetic engineering and the internet.

- Biological weapons.

Cloning

Cloning techniques are distributed according to principles and specifications of the theory and practical on several areas, genetic cloning and reproduction by nucleus. Gene cloning refers to the production of similar genes resembling the original, and the best example of this when giving birth to identical twins after a split genes of one egg and its distribution into two similar cells and grow each of them to produce a separate identical fetus. But when the cell of the embryo is sperated in the process is referred to by a particular genetic cloning, and it is growing each cell separately to produce an integrated organism researchers have successfully cloned monkeys and frogs of the cloned embryonic stem cells, similar to the original match, and it believes that it is easy to deal with embryonic stem cells because they are not discriminated by the (were not an evolutionary has turned to hair and brain and muscles and other organs) is in the wombs of their mothers, the negative aspects of cloning, the genetic test will be based in is not what will the world of him.

The third cloning occurs from mature cells, the researcher can wait to see the nature of the thing with his own eyes before they proceed on reproduction. This type of cloning is new and are using the nucleus of an adult cell adopted by Wilmot (embryologist) to clone a sheep from an adult cell, as well as it has managed to clone sheep, genetically modified with genes to produce human factors and blood clot of the most famous that carry human genes. Dolly, the most famous reproduced onganism that carry human genies while the cloned Dolly the first to reproduce the new technology. International principles for the reproduction of new life based on the cloning of a somatic cell is a mature steps each one of them have the status associated with these principles and is the following:

- The use of a cell donor (host) intended for reproduction of cells extracted from the membrane to view white pregnant sheep featuring the genes needed to form a complete sheep for the purpose of conversion to a cell capable of regeneration without preproduction and become a creature without full sexual intercourse (without vaccination) and instead of being cell embryonic viable fusion.

- Using an egg cell is extracted from sheep attached to another type of black- headed in the pot tester removed while retaining the nucleus with cytoplasm to put it right.

- The cell donors to famine and a halt to development and divisions, and preventing food resources for a period of ten days and give in to the state of sleep.

- The removed egg from the receptor cell near the nucleus of the cell donor prepared for cloning by using electric induction, small electrical firing bursts per nucleus and the egg unite and begin to act as fertilize.

- Allowing the egg genetic cells for growth and division and the formation embryo in the laboratory, implants in the uterus of sheep with black head and become the sheep that was born and was named an international replica of the white sheep of the cell donor and not the same as that used for the black bosom.

Genome

The human chromosomes is 46 consisting of strips of the double helical DNA wrapped circumvent complex shapes of helix, normal and high and consists of the DNA with four units of high repeated synthetic (nucleotides incomplete oxygen), each of which consists of three components: nitrogenous base and sugar phosphate penta and not organic.

There are four types of nitrogenous bases and symbolized by TCGA arranged in pairs along the stretch of DNA and the numbers of secondary units in the DNA molecule. Approximately 3×10^7 base pairs in each of the cells of the human body, and the length of DNA equal 8×10^3 times the distance between the earth and the moon and bigger than the distance between the earth and the sun 300 times (the length of all the DNA, In the body 2×10^{10} km).

The gene (one gene) constitutes a piece of DNA it consists of a large number of secondary units and in the molecular weight of the gene is 600×10^3 and is a very long string of four characters, and each character represents a nitrogenous base. There are usually genes in the nucleus of the cell and molecular genetic consists of two bands are linked together by special bonds, each other on some twisting spiral and there is peace on the same wrapped that consists of a sequence of nitrogenous bases, or nucleotides that contain the bases arranged in a manner different from the gene to another and then discriminate organism from the other because of all the genes governing cell functions, guidance on ways making a specific protein or another compound with medical importance, hence a single gene responsible for the general one recipe and therefore we find that the qualities beauty, shapes and colors that each one of them result of a single gene or the number of genes. Genes are transmitted from parents to offspring by mating the structural change in terms of affected and then the subsequent processes of making many of the compound causing the disease, which may be cancerous or always defect organisms.

The genome is intended to aggregate the DNA (genes) of the bacterial cell example, contain about 200- 300 gene, while the genome of a human cell include the thousand times as much as the genes found in bacterial cell 200.000 to 300.000 and the organization of these genes depends on the number, in the chromosomes of the cell Eukaryotic (human cell) is more complex than the primitive cell nucleus (bacteria).

The genetic map represents the order of genes (genes) within the cell chromosomes and that this arrangement within the human chromosomes is more complicated than other organisms. Thus, the process of discovering how to arrange these genes, given the sheer number and complexity associated with variation built and responsibility to control complex cellular functions. Hence the decoding process and diagnosis and scheduling of full human genome by genetic map and the preliminary draft of a preliminary genetic blueprint of human genes and the previous process is equivalent to a significant scientific breakthrough, scientific achievements made during the twentieth century, including the discovery of penicillin and landing on the lunar surface and use a computer and other discoveries. And according to this perception announced 26.6.2000, the end of the main phase of the Human Genome Project, which represents the first achievement in the twenty-century atheist and the development of the draft map is almost complete and a preliminary blueprint for a human gene content of the human genome and was named the human genome. It had been prepared jointly by both the research centers m the United States, Britain, Japan, France, Germany, China and other countries with long experience in genetics and genetic engineering, funded by 18 countries.

Specifications of the human genome (Genetic human map)

This map is characterized thoroughly without gaps up 99.9% and 97% of the components of the human genome, has been decoded and 85% of the sequence and gene order has been tabulated and analyzed. The rest of the map requires additional time to accomplish. The map we have opened a new era of molecular deal with the situation of life, has made it clear that in a manner distinct and different, Dr. Ahmed Zewail Nobel Laureate in nanotechnology, said that the molecules are arranged genes that act in the movement of one can not detect them, but that can be pursued with sophisticated and sensitive to femto second.

Futurism of human genetic

The discovery of the human genome and complete the approximate locations and sequencing of this large number of genes input to future developments are to:

- A new look for the human body.
- To find new ways of treating diseases (gene therapy developed) such as AIDS, cancer and heart disease.
- Correcting the genetic errors.
- Organ transplantation.
- To address the social and ethical consequences.
- Sustaining life.

The battle against cancer, AIDS and other incurable diseases are ready to locate the responsible genes for these diseases that have discovered the map and according to that can solve the problems of treating these chronic diseases and decrypt secrets, and the negative aspects of this discovery is immoral exploitation of the future such as racial discrimination in accordance with the genetic composition and control of human qualities and it requires the issuance of special legislation related to human rights and prevent the future destruction of rights (death of persons with disabilities and life of the insane, for example).

Gene therapy

Thousands of diseases due to the presence of genes responsible for the appearance and many of them dangerous to humans and non- treatment or cure and concerns and applications of biotechnology to find what is known as gene therapy, which is either by bringing the damaged gene or gene intact repair defective gene. This could be done through the intervention to repair the gene in somatic cells, or by intervening in the cell construction.

In order to spread the use of gene therapy it must be certain that the expiry date and free from damage and researchers that must be able to transport techniques and control gene expression in the correct and consistent, and should not obscure the international success of the many risks carried by this treatment.

- The genetic balance of any human being is the only thing that can not be replaced but must be preserved and transferred to the generations of while it is possible.
- Here we must make sure that it can allow gene therapy in somatic cells, and prevention must be in the manipulation because of its many negative consequences both in terms of genetic or moral.
- Gene therapy in somatic cell only affects the individual patient treat him, while affecting.
- Gene therapy in cells on the construction of successive generations.

Recent progress in the field of gene therapy

It has become possible to do some practice in the field of genetic medicine with the evolution of technology DNA "Recombinant DNA" as has been addressed most of the problems related to the production and disposal of genes and to consider their ability to modify objects, and the laboratory tests on animals proved that non- genetic, genetic medicine can be successful. It has been treated several human cells in inherited tissue laboratory for the use of retroviral vector. And proved the possibility of a peace process, the introduction of white blood that had been genetically engineered in the patient. It was also carried out several recent clinical tests for human genetic medicine, culminating in successfully treating a patient in the loss of immune complex advanced emergency resulting from the adenosine.

It seems that gene therapy is the only way to cure genetic diseases or chronic (such as cancer, Acquired Immune Deficiency Syndrome). In this case, there is an objective one is to improve the health status or save the life of a patient work of the highly desirable, and then there is no difference between the unit body and the unity of the genes. Gene therapy holds great danger is the use of this technology in order to improve the human race; can we change the balance of genetic risk for the human species? This may have dire consequences especially on the reduction of formal diversity. This method, which promises also carry the hopes of many fears. Do we have verified that all the risks in the long run, where would we be? What are our borders? Where the border between the correction of what eugenics? The achievement of this modern way of thinking must be accompanied by a profound and moral debate.

General characteristics of the human gene therapy

It was discovered more than two thousand genetic disease; all affect the genetic information in the patient and move it to the next generation. In order to restore the natural functions of the victim there are two ways in gene therapy.
- Gene therapy in somatic cell.
- Gene therapy in the cell construction.

In each of these methods a special set of scientific and ethical considerations. Construction cells are sperm cells and egg cells, which include the rest of the cells in vivo somatic cells. The gene therapy of somatic cells in the introduction of DNA in this type of cells so the added gene in the progeny of the patient on the contrary, affects gene therapy in cells on the construction rights in the early stages of embryo development.

Gene therapy in somatic cell

Before embarking on any attempt of human gene therapy it must first determine the exact mutation that leads to a particular disease. This information is not available currently, but in terms of a small number of diseases, but current advances in genetic engineering point of what will happen soon on detailed analysis of the genetic and second it should be identified on the type of mutation affected the cell in the body and their genetic transformation. Only current clinical trials to treat somatic cells, which are based on the introduction of a gene in somatic cells of a small child or a young man. Thus, cells not exposed to structural change, which prevents the transmission of the gene to offspring and this, is something which makes having a person who benefited from the treatment is always vulnerable to this disease. The structural gene therapy did not apply to humans, as it was rejected on moral grounds. But a team of researchers are currently considering the possibility of its application in the treatment of incurable genetic disease and it comes in all cases. The treatment depends on the addendum, any patient that the gene dysfunction and genetic causes of the disease will not heal or replaced, but added to the cell intact copy of it and this method of treatment do not apply except in cases of genetic diseases caused by genes elected. In general, gene expression is not valid only in a given tissue can be determined in different ways in experiments conducted on the living body, determines the quality of the input method the target tissue. Incomes through the trachea transfer of genes into the pulmonary epithelium, and injection in the liver gene transfer to liver tissue, either tumor injection in the objects is transferred genes into tumor cells also contributes to the carrier of the virus in determining the target tissue.

Treatment in the cells construction

Cells contribute to construction in the genetic heritage of successive generations. Gene therapy through the cell affect the construction of genetic stock to his descendants too, and then the sum of the genetic traits of humanity. The majority of scientists that may not be morally any attempts of this kind of treatment, another group believes that the gene therapy in the cell construction is the only way to eliminate genetic diseases suffered by millions of people.

Different methods of gene transfer

- Viruses: Retro virus, including the only to ensure that the genes transmitted via cellular divisions. We therefore consider these viruses are the most successful means of gene transfer in the laboratory, where they allow in principle a final treatment of genetic diseases.

- Chemical methods: There are numerous studies on the possible use of fat bodies and compounds multi positively charged.
- Physical methods: The target cells is ejected shells are small technisten DNA speed due to electrical discharge or explosion compressed gas.
- Intramuscular injection: Move DNA the intentions of cells around the injection site but does not merge with it, but stay for extending from a few weeks to a few months to form ring.

To deliver a specific gene or gene fragment to a cell, you should test the appropriate carrier depending on the cell type and the type of genetic defect, but for the gene replacement method, i.e. a shift in the form of direct mutated gene in position, it can not be used because most of the known vector vaccination consecutive DNA unrelated to semiconductor gives better results. For the benefit of the continuous attempts at gene therapy of the progress in the area of expanded bone marrow transplants to restore the functions of blood cells when infected with a genetic disorder, and used most of these attempts retrograde viral vectors.

New ways to design a treatment using cells of the organism
- Cell culture installed: is converted to the infected cells in organs or in the culture the appropriate gene, and then build a specific structure and in the infected tissues. Thus, cells expressing the gene and provide the onboard product withdrawal.

Diseases which are currently subjected to genetic manipulation
- Cancer: The application of gene therapy in cancer is not now aims to correct genetic defects in somatic cells, but sought to allow the introduction of genes to eliminate them.

- Neurological disease: This treatment is used to reduce nerve damage that accompanies Parkinson's and Alzheimer's disease and to enable the infected neurons to recover their function.

- Acquired Immune Deficiency Syndrome: There are many techniques under the anti- vaccination testing, such as installation and self- inoculation cultures for the primary fiber cells that contain transience's carry code of viral proteins and immunization procedure by genes which are to block viral reproduction.

- Gene therapy of developed AIDS, the first attempt of gene therapy in humans (1990), while the enzyme that remove amin of adenosine make the child is not capable of performing the immune response to resist infections after isolation of lymphocytes from patients, and then insert a normal gent for an enzyme that remove the amine through virus vector.

The treatment of patients through the provision of the necessary genes are still the idea compelling. For those who remain in front of researchers in basic science and conscience, much to be done before gene therapy to succeed.

Fingerprint

That science is progressing dramatically in the current year, so that it can be made in the last quarter of the last century, equivalent to human progress in its long history as a whole. In the field of genetics offers this impressive progress of science and builds on the many hopes in the future of human. While the human in a state of surprise and astonishment, which inherited the technology to adapt the results of gene. The scientists that discover some of the problems to show us the genes which was later named the fingerprint genetic fingerprinting. What are these? What are the issues that can be resolved, and is unable to traditional means of forensic medicine to find a solution? Genes that carry genetic message from one generation to another and guide the activities of each cell is a giant molecule may be converted to resemble the strings, called DNA that contains all the genetic traits, from eye color to the smallest structures in the body. They are the result of genetics in human cells, for 23 pairs of chromosomes in the nucleus of the cell, the chromosomes constitute from DNA. Proteins play an important role in maintaining the structure of the genetic material that lead to reveal the full individual. The finger printing began until 1984, when the geneticist deploy at the University of Leicester in London search of genetic material that may be repeated several times and re- sequences itself incomprehensible represented in the length and location, this research reach in one year that this sequences characteristic of each individual and can not be similar between the two except in cases of identical twins only, and the potential similarity of fingerprints between one person and another one trillion, making it impossible similarity, that was found that these differences are unique to each person just like fingerprints and therefore called the fingerprint genes. Dr. Alec has recorded his discovery in 1985 and named them the name of the sequences of the human person as defined and as a means of identifying a person through the passages approach sometimes called DNA fingerprint "DNA Typing". The genetic fingerprint known through the courts, although had spent time in the detection through forensic medicine, where possible knowledge of this fingerprint to identify the mutilated bodies and tracking children and missing soldiers, as it can mark genes to identify the person until the bulbs of the hair that has been cleared of many of the defendants by identifying the genetic fingerprint of murder, rape and revealed the true perpetrator of the crime, had a genetic fingerprint of the word on the issue of genealogy polarities of a number of issues to prove paternity, rape, and calculates the ratio of the distinction between individuals using fingerprint genes found that this ratio up to about 1: 300 million people, there is one person with the same genetic fingerprint was also found that fingerprint genes inherited according to Mendel's laws of genetics. It has also found that the fingerprint genes vary according to geographical patterns of the genes in the peoples of the world, for example, is different from Asian (Mongolian or yellow race) for the Africans. For the identification of genetic fingerprint requires a small sample of tissue that can be drawn DNA including, for example:

- A sample of blood in the case to prove filiation.
- A sample of sperm in the case of rape.
- A piece of skin under the nails or hair roots of the body in case of death after resisting the aggressor.
- Blood or semen frozen or dry is on the crime scene.

- A sample of saliva.

Ethical implications

That this issue is distinct attention to enrich the scientific research related to this section of the scientific specialization, which is still growing and evolving as a number of questions that put precision together constitute the social issues and scientific issues that require a unique answer response send in self- certainty and a sense of security and assurance. The religions have confirmed the ethics of researcher and research ethics and both sides of the same coin, a search should be moving to the reconstruction and development and preservation of the environment that God created it so well, the search if deviated from their destination and good career development research is not useful and must be liberated with the production and consumption together.

The importance of this subject is first that it does not affect the religious, but very cautiously and in accordance with insights and analysis are limited, and the cure of genetic testing and abortion, infertility and human eugenics and other topics related to the needs of the Muslim scholars to discuss and study and comparison with the fundamentals of the faith and purposes of the law. If we were not the courage and wisdom to show the religious scientific opinion on these issues inherent in our daily lives, will remain controversial among the various currents and contradictory beliefs, which reflects negatively on our future generations and directly affects our faith, one way or another.

Ethical considerations

The gene therapy in somatic cells aims to treat serious diseases, and the possibility of morally acceptable. The gene therapy in cells construction remains a subject of controversy with regard to cell construction cells and with regard there are a number of questions.

- Do we have the right to change the genome of an unborn child?
- Who has the right to approve?
- Are we encouraging the introduction of genes (such as growth hormone) to improve the quality of embryos? Any non-therapeutic uses.

Despite the many considerations of discussions on gene therapy technology, millions of people with one of the different types of genetic disease, they hope to apply these technological developments soon in the attempts to mankind and in the absence of other types of treatment, then should allow the growth and development of gene therapy in somatic cells, under the supervision of bio-security. Which include preventive measures that should be adopted to reduce the need for gene therapy in Muslim societies:

- Interests in genetic counseling in public and private hospitals to help people to absorb health education on genetic diseases and to take the necessary measures.
- Promoting genetic studies (epidemiological) in families and tribes that carry infected gene and this makes it easier for genetic counseling, as well as gene therapy.
- Do not marry relatives, particularly when it is in the family ancestors are infected with diseases and hereditary.

Medical consideration

It is not preferred to attempts to human gene therapy in the absence of a broad scientific background able to understand the nature of genetic disease and molecular consequences. On the other hand it must be used human gene therapy techniques in the framework of a particular lead to unwanted hard impact and to restore normal cellular function in a person's life that continue throughout the future. It also must be gene expression regulated outside the original a manner as to improve the patient's condition without damaging the cells or the person the future.

The use of viral vectors in human gene therapy is critical concern due to the ability of these vectors to the initiate the particles conditions infected virus that may spread to neighboring cells, or to others in the community.

The treatment of structural cell may cause damage that occur in future generations, and may lead to correct the composition of the affected gene mutations. The remaining operations targeting stay primitive and with non-controlled roads.

Security considerations

There are potential dangers from the use of gene transfer by retrogressive virus but it did not cause any minor damage in humans. The National Institutes of Health has described in the United States malignant T cells in monkeys, but discovered later that these resulted from contamination of the carrier virus.

Religious considerations

- God created man in the best stature and with generosity to other creatures, and tampering with components of the human being and subjected to tests of genetic engineering without a goal is incompatible with the dignity that God bestowed on humans will read on him, "We have honored the sons of Adam".

- Islam is a religion of science and knowledge as stated in the verse, "Are those equal who know and those who do not know", which is not forbidden to the human mind in the field of scientific research and useful genetic engineering in its various aspects in addition to knowledge.

- Everyone has the right to respect dignity and rights whatever genetic characteristics.

- It not conduct any research or carry out any treatment or diagnosis of the genome, of any person unless conducting rigorous and prior assessment of dangers and potential benefits associated with these activities with a commitment to the provisions of law and ethics of this matter and, if not beneficial to health and direct benefit to him. It should respect the right of every person to decide whether he wants or does not want to take note of the results of any examination or genetic consequences.

- All genetic diagnoses, preservation or preparation for the purposes of scientific research or for any genetic examination or its consequences are confidential.

- It is not permissible to offer any person for any form of discrimination based on genetic characteristics, which shall be liable or result of reducing the fundamental rights and freedoms and violating the dignity.

- No research on the human genome or any of this research, particularly in the fields of biology, genetics and medicine, should prevail over the observance of human rights and fundamental freedoms and human dignity of any individual or group of individuals.

- Publication of books should be to simplify scientific information about genetics and genetic engineering to raise awareness and strengthening on the subject.
- The introduction of genetic engineering into the curriculum at different stages and in local media.

Futures Genetic Engineering and Biotechnology and the Internet

Many believe that one of the most important early developments that will emerge from genetic engineering is the technology of artificial viruses, that could become almost as today's design and manufacture of cells and viruses that have changed and stop specific biological processes. Thereby eliminating a particular disease or changes in the characteristics of an individual. Biologists and researchers hope to be making a virus able to recognize the cancer cells and access to arrest its proliferation as well as many other applications and thus allow the virus to replace the surgeon industrial tools and unequivocal chemotherapy drugs, the strongest and the most disturbing and less harmful in terms of side effects.

On the basis of current developments in science and technology which they can draw a picture of a bright future for humanity that can be created by itself if it wanted worked according to ethical, social- science concepts. Then will come aduy on humanity as a whole interconnected network giant relies on a large group of small satellites for communications and electricity will be available in areas. Thanks to remote farms with genetically engineered to convert sunlight into carbon and then to the raw stream and can then run all the equipment and facilities, including communication devices via satellite and the Internet. It had been predicted by many scholars the most important scientific developments of civilization that have been achieved, including information technology and the Internet, artificial intelligence, and the travel space with full visualization and beautiful and optimistic to the future of woven colored threads of science, ethics and technology, philosophy and focus on the short term more than the remote for the application of the Scriptures space because of the failures observed in the draft and the space agencies of U.S. and Russian (Mir station in particular) as well as matter in relation to artificial intelligence.

The biggest and most important feature in information technology and new communication to overcome language barriers and ignored the local culture and traditions and there is no technical obstacles to prevent them from connecting the world and its peoples to each other. But the potential for the delivery of information today has become much easier than the capacity of countries to deliver water and electricity and the provision of medical and housing for their people. Of course, the Internet can not solve all the problems of the world's social and economic development, but we started to see positive effects in many areas and a variety of no conceivable when one plans for the network. It is a positive revolution which imposed itself.

To illustrate this category which used to computers and networks of those who daily through the Internet and its information online and link part of their careers and those already receiving significant superiority to the other categories and who update the world via the Internet are in transition to a new server- class gap between the connectors are growing rapidly.

Either with regard to biotechnology and genetic engineering then it is of course quite different, the world has witnessed rapid progress and as a surprise to us and put the sheep reproduce and international human genome project at the forefront of current scientific. In our vision will bring us a greater developments and surprises in our lives from the Internet and genetic engineering, especially after the human genome and not from the sun or space.

Two recent examples will come believes to many of the biggest scientific surprises that have occurred over the last few years, first cloned sheep and the international human genome in front of the computer any more than artificial intelligence to human intelligence for the first time. Some people view the prospect of human cloning possibilities and tremendous results and violating some positive, some seriously, in the social and cultural heritage will be able to parents as soon say- the possibility of using cloning technology and genetic engineering to compensate for specific genes to their children before their training. This will change the capacity of children's physical and mental to protect from certain diseases and symptoms and arming the other capabilities to facilitate their life around them. But this technology in the first decades at least will be expensive and most likely this will lead to expansion of the difference between the two layers of rich human or genetically grafted and natural. There is no doubt that this will push the old division of mankind into masters and slaves, unless you turned this technology accessible to everyone, and this is expected.

Protein map "Proteome"

The term "Proteome" which appeared in 1994 to the total pool of proteins present in each cell type the amount of a hundred trillion in each individual and the total proteins produced by cells of the body during different life stages.

After discovering the human genome, which includes (full content of genes (genes) in the amount of the 34 thousands people only, and not one hundred thousand. I think scientists for a long time) and also all the genes inherent in the cells of the body at the present time highlights the important question, what the protein content of these cells.

The type of each protein has to be known as a result of these cells and what function each protein and then what order of these proteins. Asking this question came after attrition rationale the concept of the genome and its consumption and is not enough to know as responsible for stimulating cells to produce the kinds of protein, but only requires the knowledge of the situation in its entirety in routine cases of disease and natural and in accordance with these questions and answers on the back of proteome.

Proteome contains information more complicated and the secrets of the genome is more dangerous than those found in the genome and extensive knowledge and synthetic for more than a million different types of proteins. The concept of proteome is known later human proteome is doing now by scientists and they hope that these will be the beginning of the main achievements under this project, despite the severe difficulties faced by these scientists, in excess of those related to the human genome. The analysis (cell proteome), reached some of the researchers in 2000 to build automated device called the molecular scanner "Molecular Scanner", which is carried out by measuring by mass spectrometer, from which tens of thousands of known proteins in a single day, and at the speed of more than ten times what was known before.

These researchers also managed to build a million boosted the analysis of protein per day to build bigger infrastructure database proteomics mankind. The draft of human proteome or other whereupon many of the laboratories and big budgets and international companies such as the famous, hybrgenics,

Clera Genomio, different research directions, the analysis of three-dimensional protein structure and interactions between proteins, which performed many of the key characteristics of the human proteome would pay off represented by the following:

- Specification of fungus or yeast proteome with a single- cell, the first that has been done in the world of proteome.
- This project change from how to design drugs in the near future.
- The appearance of the so- called science and technology human proteome, which will focus on the conversion of most of the drugs manufactured by genetic engineering and biotechnology.

Genetically Modified organisms (GMO)

Biotechnology, which contains the processes of nucleic acid technology and molecular biology to separate the specific gene from one organism and transferred to a particular object of another district called gene transfer technology, "Transgenic" or may be called the genetic change or genetic modification "Gene modification" and called on the living modified organisms has been applied this technique recently on agricultural crops in the recent developments in genetics, which is also hot topics in it. A number of genetic modifications on some common food organisms, the addition of a specific gene or several genes, for example when carrying out genetic modification of wheat plant is usually a small percentage due to the fact that this plant has about 80.000 genes. The process of genetic modification is possible in practice so as not to become genetically modified plant to another object or to plant malicious, but maintains the general attributes of the amendment with relative injury. According to some voices of opposition to the process of genetic modification, that could cause damage to humans and the environment, including poisoning or allergies.

The number of countries including the United States of America, Canada and China that will produce genetically modified crops, including soybeans, corn, flax, beets, potatoes in different proportions. From a technical point of view alone, there are a number of benefits of genetic modification of crops to convert to regular crops resistant to pesticides and weed, disease and insects and reduce pesticide use and increase productivity and improve the nutritional value of crops, and make it more a shift of the circumstances, including salinity, drought and an increase in the quality of the crops for use in food as well as in withstand the transport and storage and make crops resistant to pesticides, insects or insect resistance or both groups. The genetic modification was still in use in plant breeding has a significant impact in providing food for humans and methods that have been used traditionally to improve the crops, but they are not specific or accurate results of modern genetic modification in which the change is unknown in most cases, things such as crops and breeding plant breeding and mutations.

Bio- safety of genetically modified organisms

The international convention indicate on genetically modified organisms to the need for special tests on GMOs for fear of the potential impacts on human health and the environment of these tests is the requirements of the United Nations Environment Program and the Organization for Economic Cooperation and Development.

The bio- safety grounds due to the danger potential that result from using this technology mismatch impact on human health, animals and impact on the environment, there is the potential transfer of toxic compounds from one object to another or create new toxins and stresses the Food and Agriculture Organization of the necessity of conducting these tests and countries of this technique. Voices opposition to genetically modified organisms after the production of genetically modified crops for food within a specific industry, voices opposition to it were carried out adoption these voices opposition to genetic modification of agricultural crops on a number of reasons, including:

- Exaggeration to talk about the bio- safety of agricultural crops genetically modified organisms.
- Bias the producers of GM crops.
- An evaluation of the safety of this vital crop locally in the producing countries.
- The introduction of these foods in the world trade conflicts because the majority of exporters are the countries that grow these crops.
- Evaluation of bio- safety in the importing countries.
- Lack of uniformity in laws relating to the products of such countries that imported and produced.

The positions of countries and international bodies of genetically modified crops countries vary in their positions on genetically modified crops. Some producers and other consumers, but they all agree on the specification is limited information on the food label of the product of GM. The European Union is marketing a number of modified food while China and Australia are sowing genetically modified crops and Japan imported crops and genetically modified food and shopping, many of which, while South Africa imports some food container material from genetically modified crops.

International standards for the product of genetically modified

There is much debate about the GM product and how to dispose of it in terms of safety and health, and where the consumer's right to choose. The EU believes that the consumer's right to know the chemical composition of food, depending on the nature of the product. Protein must know that while the components for oil and sugar is not necessary with the definition of the product. It is worth mentioning that there was a project for an international standard developed by the international body that develops the specifications for the food. Adoption this standard on the principle of similar semi- finished between GM and the current food intake is limited and there is no justification to identify him and did not know that there must be full use of the terms of use, installation, and the source.

Advanced technologies

That the era of advanced technologies "High Technologies" or high-technology "Super Technologies" in which we live the last three decades of the twentieth century, the era in which we do not know how many decades it will take, representing a number of scientific areas and new technology comes on top of these technologies, laser and fiber- optic and space technology, new materials, pharmaceuticals, chemicals, minute nanotechnology, and finally biotechnology and genetic engineering.

The forthcoming technical applications that are difficult to know the extent of today and its impact on humanity can be viewed as the era of advanced technologies as the following day when mankind as a whole interconnected network giant relies on a wide range of communications satellites such as radio waves and X- ray laser, so that every part of the ground contact one of the satellites in the moment and will be available electricity in remote areas with farms, genetically engineered to convert sunlight into carbon and then to the crude stream, and can then run all the equipment and facilities, communications equipment, including satellite and the Internet.

The future applications of these technologies will be radical changes in the forms of life activities and practices relevant to the interests of individuals, groups and the process of coordination between these advanced technologies is a strategic way to bring about a surge in operations research and industrial beginnings began to appear, for example, a draft genome and bioinformatics. We will try in this article and subsequent articles offer examples of advanced technologies.

Femto

It means the number 15 and the chemistry of femto, to understand the reasons that lead to some chemical reactions without the other, one of the achievements made at the end of the twentieth century and the efforts of the world that have emerged Ahmed Zewail, who won the Nobel Prize in Chemistry in 1999 and showing the possibility of seeing how to move the atoms within molecules during chemical reactions using laser technology and the rapid use of a new standard of time is Alfmto seconds (10^{15} seconds). Zewail has been used pulses of laser beam of a partial vacuum in the middle of materials to study the chemistry of high- speed stages of the transition, working within the Alfmto seconds be managed after the suddenness of molecules in the interim period and then became a pioneer of so- called Alfmto chemistry using laser technology (laser Alfmto) camera and a very fast, sophisticated and very accurate to portray the ongoing chemical reaction between the molecules in three- dimensional image Alfmto time in its three dimensions, not one dimension only.

Finally, what scientist do is to identify cases of transition of chemical reactions as broken links and new links up, and the development of new chemistry carried the name Alfmto result of invention, or a new laser called laser Alfmto or laser technology and through rapid as we were filmed for the moment the chemical reaction within the atoms in the process of only one part of a billion a second, and therefore this technology and its owner, Dr.

Zewail laser secrets complex world characterized by inventing something new the properties of new energy and knowledge of the movement of particles from birth or docking to know what was happening in record time is a million billionth of a second the proportion of this period to the second equivalent of one per second span of time to 22 million years ago.

To reach Dr. Ahmed Zewail of the use of laser microscopy to clarify the picture may have been the most difficult times in less than two and thanks to the time factor has been developed to see things, whether internal or external speed and one millionth of a billionth of a second.

The features of Applied Chemistry Alfmto side is represented as medical, industrial and agricultural in nature and changes in the human body, such as treatment of diseases such as cancer, diabetes, a cell can be imaged in the human body, and according to that disease can be determined in the light of the nature of these cells. That laser Alfmto which has been utilized for imaging the moment of the chemical reaction within the atoms in the cell process is not only is one of the thousand billion from the second.

Nanotechnology

nanotechnology, the technology that deals nanometer scale (metric unit of measurement), which is manipulating atoms for the manufacture of automatic equipment and information does not extend far a handful of atoms, and then anything can be made by micro- physics is quite different. The first part of the term to reflect the unit of measure (nano= 190^9 meters)

the term nanotechnology in Engines of Creation written in 1986 and said the possibility of seeing the future of the armies of machinery hidden carry oxygen and nutrients and waste, and manufacture of atomic- sized machines called complexes pads that hold individual atoms. It was found that the control of the maize one and move freely and easily from attributes of nanotechnology.

This technique showed high density in the form of recent innovations in many of the global scientific publications. Including the mutations responsible for many genetic diseases and therefore provide in the future and relevance of information is indicated task to determine the organism is an information system for the manufacture of proteins and other compounds in spite of the inability to resolve the issue of structural triangular shape of proteins, despite the existence of mathematical models serve the purpose.

A fruitful field of nanotechnology research in many parts of the world and in government labs, commercial and academic, have emerged, according to the products on this technique such as sewing pants of fiber and manufacture of precision tennis balls retain flexibility. In the near future, computers will appear smaller tubes made of carbon atom chips represent the atomic scale wires and high strength to build elevator to space, and plants that will manufacture computers minute integrated directly with the human brain to increase intelligence. It is clear from the examples mentioned that nanotechnology is a technology that will change very little minutes every aspect of human life and giving people the ability to control the material, and this technique is the most important applications of medical treatment of human beings through the introduction of precision instruments within the cells to repair infected objects from within or for the diagnosis of patients as well as some developments on the mechanisms of control cells. The first medical use of this technology has been developed a device implanted in the body that may manufacture.

The energy comes biofuels in the cell and the purpose of this engine is the integration of machines in living systems fully, and use some of the scholars of this technology to produce nano- bombs to kill cancer cells. A team of other customers of this technology to produce nano- bombs to kill cancer cells. A team of scientists of any other industry, the crew of siliceous teeth not larger than the size of the cell that can swallow the red blood cells and re-launched into the bloodstream, either antibiotics nano-particles "Nano biotics", which new types of antibiotics contribute to solving the problem of resistance of some types of bacteria to drugs, as well as the modified bacteria organisms, are converging nano- tube micro- rings 2.5 nm diameter amino acids and small hole walls of bacteria are infectious.

Researchers believe that the future of medicine is moving towards nanotechnology that will change medicine, as future devices that will work within the human body to diagnose many diseases and treatment. Russian scientists in the field of quantitative light and laser physics to reach a new discovery has been called the needles agency, a new type of X- ray beam, or special characteristics, as containing the elements of nano- any electronic material on subatomic particles that do not exceed the measurements of nanometer dimensions . It has also developed the first computer chip companies auction that could contribute to increase the power of computers and a reduction, while reducing the amount of energy consumed by the chip is composed of cylindrical molecules of carbon atoms in diameter than a billion to a part of the linker carbon (smaller than a hair a hundred thousand times).

Diagnostic

Securing the different types of health services, preventive, curative and rehabilitation of the basic necessities of the individual and society is part of the economic and social development, the Ministry of Health before the embargo the country and the terms of reference of modern medical equipment and refurbished medical equipment, such that the health services in the country in all its aspects to the stage of qualitative and quantitative development admittedly many of the specialized agencies and international experts. That the imposition of the embargo has negatively impacted on the level of health services and spare the necessary medical supplies such as vaccines, medicines and laboratory solutions. In spite of that medicines and medical supplies but not prohibited under UN resolutions, but that the need for medicines and medical supplies that have increased due to the deteriorating state of health, environmental and food, which led to the emergence of many diseases, and chronic diseases and malnutrition. For example, Iraq was clean and free of cholera; the disease returned and appeared again significantly in 1991. The scarcity of essential drugs and lack of availability of the required quantities led to the deterioration of the situation of citizens suffering from chronic diseases such as sugar and heart disease, hypertension, epilepsy, kidney failure and cancer diseases.

The laboratory tests were not no better than drugs, because the lack of laboratory materials and equipment used to conduct those tests and lack of maintenance and sustaining them available because of the acute shortage of spare parts and failing to be delivered to Iraq as well as the lack of diagnostic kits necessary to conduct examinations and laboratory tests, all that reflected negatively on the number of tests performed annually following table shows the percentage decline in the monthly Madal to prepare laboratory tests compared to 1989.

Impediments to the implementation of the diagnostic kits (negatives)

- The lack of some raw materials necessary for the completion of diagnostic kits, including:
  - Chemicals.
  - Other essentials.
  - Hardware.
- Difficulties in meeting the needs of the researcher:
  - Chemicals.
  - Services in the local market.
- Continuing attempts to obtain materials and devices from other outlets outside the country led to:
  - The survival of the need for quite a few of the resources vernacular.
- The cost of scientific resources, stationery and print the necessary reports to parents of high.
  - The difficulty in obtaining journals and literature of modern world.
- The area of examination and evaluation
  - Assigning one for the purpose of examination and evaluation.
  - Not possible to give a certificate of inspection for some diagnostic kits for the following reasons the amount of material sent for testing are limited, instabilities of some materials, lack of some modern techniques and equipment.
  - Unable to implement a number of these numbers.
- A steady increase in prices of materials and equipment and the cost of sustaining an impact on services:
  - Estimates of the prices offered by the researchers.
  - A number of research and is now paragraph of materials and devices is the amount greater than what they have as much time of signing the contract.
- Lack of standard materials and solutions for a number of diagnostic as reference material for the purpose of comparison scanned materials and productive and the lack of number of standard delay in conducting the tests or they can not be implemented with the refernces.

Technology and Science

Future Studied are focusing on several scientific areas such as renewable energy, genetic engineering, biotechnology, electronic industries, the manufacture of computers and communications, telecommunications space and material science. Researchers expect, the use of satellites for the transfer of solar energy to micro-wave stations can broadcast to the ground as the receipt and then transformed again into energy that can be used, then in the field of genetic engineering, many recalled of the perceptions of future scientists in the following futurism:

- Copying Genius reproduction free of disease.
- Production that produces roots potato tubers, while spared the same plant tomatoes.
- Production of human beings does not depend on food animals, but on solar energy and carbon dioxide since it contains chlorophyll.

In the area of electronics and computers industry witnessed the following:

- Computers that help the doctors in the future to conduct the necessary tests, such as analysis of blood and others, then diagnosis of disease and provide medicine for patients.
- The management of entire houses by computers, especially those inhabited by people with disabilities.

In transportation and communications, the human beings will benefit in the future from doing the following:

- Doing his own research without going to the university, but even before his computer.
- Attending the scientific conferences held thousands of kilometers away and participates in the discussions without the need to be present physically.
- Weather forecasting, prediction riches buried within the earth.

In the area of space, scientists also forecast to do great achievements at the level of outer space, including:

- The establishment of settlements in space, the moon and Mars, and the atmosphere surrounding land.
- Establishing satellite factories producing many electronic components and medicines.
- Conquest of outer space and freezing of embryos then placed them on the spacecraft, which will pass through space.

In the future, the peoples will be able to participate in the materials industry to produce materials with distinct qualities that can be used in the industry of cars, garment, as well as the use of carbon instead of silicon in the manufacture of bio-computers, and in other areas, there will be success in the war against the diseases, scientists expect that they will know the secret of the cancer, then to destroy it, also to understand the old age, and then to find the means to prolong life.

In about two hundred years, there will be four possible scenarios for future reference, namely they are:
- Very pessimistic.
- Cautiously pessimistic.
- Cautiously optimistic.
- Eager for growth and technology.

In future new generations will be created and involved, the so-called future shock and Toffler goes through analysis that includes:
- The future shock is a severe illness that affected increasing numbers of human beings and can be called satisfactory inability to adopt to rapid change.
- Reactions to the future depends on what is known about the ability to adapt recalled hypothesis in the 1st. instance, that the proliferation of psychiatric and neurological a lot of people due to the shock of the future and technological progress and scientific facts have not been able to create awareness on assimilations.

References

- Lewine, M.A., congnitive theory of learning, 1973, N.Y. Wiley.
- Robert M., Gane, Essentials of learning for instruction, 1975, Holt Reinehort and Winston.
- Travers, essential of learning, 1997, N.Y. Macmman.
- Borich, Gary D.: Effective teaching methods. Second edition, 1992, Macmilan.
- Vaizey, J., and Sheehan, J., Resources for education, 1968, Allen and UNWIN.
- Andrew Taylor, Frances Hill "Quality management in education" in (organization effectiveness and improvement in education), edited by Alma Harris 1997, open university, press, Backing ham.
- Jerry Banks: Principles of quality control, 1989, John Wiley: sons, N.Y.
- AECT, Educational Technology: Definition and Glossary of Terms, Vol. 1, 1977, Vol. 1. Washington, DC: AECT.
- Mackenzie, Norman and others: open learning systems and problems in post- secondary education, 1975, Paris: Unesro press.
- Lenrer. J (2000): learning disabilities theories diagnosis and teaching diagnosis (6th ed) (2000), Houghton Miffin Co. Boston.
- Mercer, G.: students with learning disabilities (1997), Columbus OH prentice- Hall INC.
- Polloway, EA: Patton. J.: strategies for teaching learners with special. (1993), Merrill – Newjersy, Columbus- Ohio.

- Boyd. Gary: the shapping of educational technology by cultural politics and vice versa. Educational and training technology international (1991), 28(2).
- American Association for advancement of science, science for all Americans (1989), project 2661, Washington, D.C.
- Anthony D.F. and Dean L.C. science for all children elementary school methods, (1998) MSA, Waveland press.
- Croline, Mc. G. science: technology and science (1999). Handbook, the association for science education.
- Chun, S., Scientific literacy an educational goal of the past two costing NARST Annual Meeting (1999), Boston, Massachusetts.
- Korla, R.M.: popularizing science in schools (2000), Delhi. Ram Prmto-graph.
- Encyclopedia Americana, Vol, 15, Japan: (1981), education.
- Carl Parsons Quality improvement in education, (1994), David Fulton Publisher Ltd- London.
- John Dewy, philosophy of education (1985) Littlefield Adams and Co. N.Y.
- Correa H: A national Methods in educational Planning and Administration, (1979), N.Y. David Mcky.
- Bozeman, W.C. Educational Technology (1995) best practices from American school, N.Y. Eye on education.
- Ellington, Henry, and others, Handbook of educational technology (1995), Kongan page, 3rd edition.
- Gagne, R.,: The conditions of learning (1985), New York, Holt, Rinehart and Winston.
- Garadner, H., Creating minds, (1993), New York.

.Al- Saadi, 2, 5, 2006: Iraq's National vision, strategy, and policies: strategic insights, vol. v.

. Baker, R. 2004: Iraq and human development: culture, education and the globalization of hope.

.Bhagwati, Jaqdish 1979: International Migration of the Highly skilled: Economics Ethics and Taxes, Third world Quaterly Vol.1 No. 3, July, pp. 17 – 21/ 1976.

.Bhagwati, J. and Parington, M. (eds) 1976: Taxing the Brain Drain, 2 Vols, Amsterdam: North Holland Pub. LO.; New York, American Elsevier Publishing.

.Davey, M. E. 2003. Federal research and development funding: FY 2004. Congressional Research Service, the library of congress, order code IB 10117.

. European Commission – Directorate General for Research, key figures 2003 -2004, towards a European research area: science, technology, & innovation, p 23.

.Fossum, D., Painter, L.S., Eisman, E., Ettedgui, e. and Adamson, D.M. 2004. Vital Assets: Federal Investment in Research and Development at the Nation's Universities and Colleges. Rand Corporation, Santa Monica, CA.

.Ghali, H. 2005: The destruction of Iraq's Educational system under us occupation, center for research on Globalisations.

.International Education standard for professional accounts 7 (2004): continuing development: A program of lifelong learning and continuing development of professional competence.

.National academy of Science. 1997, preparing for the 21st century: Science and engineering research in a changing world. National Academy of science. 12p.

. National Institutes of Health. 2004, Summary of the FY 2005 president's budget. 12p.

.National Science Foundation b. 2005, The budget for fiscal year 2005. National Science Foundation, 2002. Survey of R &D Expenditures at Universities & Colleges, Fiscal Year 2002.

.Pharmaceutical Industry Profile. 2003, Dramatic growth of research and development, (chapter 2): pp 10-22,.

.Qasim Subhi. 1998, Research and development in the Arab States. Development of S&T Indicators, UNESCO, (Cairo) ESCWA.

.Science Watch. 2003, Middle Eastern Nations Making Their Mark, Vol.14, No.6,.

.UNESCO Institute for Statistics. 2003, Immediate, Medium and Long – term Strategy in Science and Technology Statistics. UNESCO Institute for Statistics, Montreal.

.UNESCO Institute for Statistics. 2004, A decade of investment in research and development (R&D): 1990-2000. UIS Bulletin on Science and Technology Statistics. Issue No.1:1-4.

.Iraq 2004: Iraq, Education overview.

.Word Education services – Canada (WES), WEP- Iraq, 2004: Iraq Higher Education.

Prof.Dr. Sami Al-Mudhaffar FIAS,IAS

- Has obtained a PhD Degree in Biochemistry in 1967..
- He was posted in 1967 on teaching and research assignment at the College 0f science, University of Basrah, then promoted to assistant Professor, and in 1979 to Professor of Biochemistry at Baghdad University.
- He has played an modest role in promoting Biochemistry and related subjects such as Molecular Biotechnology and clinical endocrinology and enzymology research and was mainly responsible for initiating and establishing many scientific and educational bilateral programmes with scientific organizations and laboratories of the advanced world.
- From 1968 till 2003, was lecturing to undergraduate and post graduates students at the college of science and other colleges of University of Baghdad and other Universities
- He was a scientific supervisor of 30 PhDs and 150 MSc students in the field of Biochemistry and related subjects
- Published more than 50 inventions and 270 scientific papers in various diseases .

www.ingramcontent.com/pod-product-compliance
Lightning Source LLC
Chambersburg PA
CBHW081138180526
45170CB00006B/1848